INTRODUCTION TO
MODERN ALGEBRA

INTRODUCTION TO
MODERN ALGEBRA

– *Author* –
Anuj Gupta
Parshant Kumar

Wordpen Academics
Publishing Science That Matter

ISBN: 978-81-19283-22-4 (Hardback)

Published by : **WORDPEN ACADEMICS**
4736/23, Ansari Road, Darya Ganj,
New Delhi – 110 002
Phone: +91-011-35004336
E-mail: info@wordpenacademics.com

Printed at : **Replika Press Pvt. Ltd.**

Exclusive Distributor: ***Bio-Green Books, New Delhi***

PREFACE

Modern algebra, also known as abstract algebra, is a branch of mathematics that studies algebraic structures, which are sets of elements with certain operations defined on them. It is a powerful tool for solving complex problems in mathematics, physics, computer science, and many other fields. The history of modern algebra dates back to the early 19th century, when mathematicians like Niels Henrik Abel and Evariste Galois developed the theory of equations and group theory. These early developments paved the way for modern algebra, which emerged in the late 19th century with the work of mathematicians like Georg Cantor, Richard Dedekind, and David Hilbert. Modern algebra encompasses several branches of algebra, including group theory, ring theory, field theory, and module theory. Each of these branches studies algebraic structures with different properties and operations.

Group theory is the study of groups, which are sets of elements with an operation that satisfies certain axioms, such as closure, associativity, and the existence of an identity and inverse element. Group theory has applications in many areas, including cryptography, physics, and chemistry. Ring theory is the study of rings, which are sets of elements with two operations, addition and multiplication, that satisfy certain axioms, such as associativity, distributivity, and the existence of an identity and inverse element under addition. Ring theory has applications in number theory, algebraic geometry, and coding theory. Field theory is the study of fields, which are sets of elements with two operations, addition and multiplication, that satisfy certain axioms, such as associativity, distributivity, and the existence of an identity and inverse element under both operations. Field theory has applications in algebraic geometry, number theory, and physics. Module theory is the study of modules, which are generalizations of vector spaces over a field. Modules have applications in algebraic topology, representation theory, and algebraic geometry.

Modern algebra has played a significant role in the development of many areas of mathematics and science. It has provided powerful tools for solving complex problems and has led to the discovery of new mathematical structures and relationships. Modern algebra has also led to the development of new fields of mathematics, such as category theory and homological algebra, which have had significant impacts on other areas of mathematics and science. Modern algebra is a fascinating and powerful branch of mathematics with many applications in various fields of science and engineering. Its theoretical foundations and practical applications have made it an indispensable tool for solving complex problems and exploring new mathematical structures and relationships.

It covers a range of topics related to the study of algebra, from the basics of set theory to advanced topics such as Galois Theory and representation theory. The book begins with an overview of modern algebra and its importance in various fields of mathematics and also covers the history and development of ring theory, including its applications in algebraic geometry and number theory. The book covers the basics of set theory, including the axioms of set theory and cardinality and introduces module theory, which is a generalization of vector spaces, and its applications in various fields of mathematics. The book discusses the applications of commutative algebra in algebraic geometry and algebraic number theory and also includes a discussion of current research topics in group theory. The book covers the history and development of Galois Theory and discusses the importance of representation theory in various fields of mathematics, including physics and geometry. It is suitable for undergraduate and graduate students as well as researchers in mathematics and related fields.

We are grateful to all those persons as well as various books, manuals, periodicals, magazines, journals etc. that helped in the preparation of this book. In spite of the best efforts, it is possible that some errors may have occurred into the compilation and editing of the book. Further queries, constructive suggestions and criticisms for the improvement of the book are always welcomed and shall be thankfully acknowledged.

Anuj Gupta

Parshant Kumar

CONTENTS

1

MODERN ALGEBRA: OVERVIEW

Modern algebra, also called abstract algebra, branch of mathematics concerned with the general algebraic structure of various sets (such as real numbers, complex numbers, matrices, and vector spaces), rather than rules and procedures for manipulating their individual elements.

During the second half of the 19th century, various important mathematical advances led to the study of sets in which any two elements can be added or multiplied together to give a third element of the same set. The elements of the sets concerned could be numbers, functions, or some other objects. As the techniques involved were similar, it seemed reasonable to consider the sets, rather than their elements, to be the objects of primary concern. A definitive treatise, Modern Algebra, was written in 1930 by the Dutch mathematician Bartel van der Waerden, and the subject has had a deep effect on almost every branch of mathematics.

In algebra, which is a broad division of mathematics, abstract algebra is the study of algebraic structures. Algebraic structures include groups, rings, fields, modules, vector spaces, lattices, and algebras. The term abstract algebra was coined in the early 20th century to distinguish this area of study from the other parts of algebra.Algebraic structures, with their associated homomor phisms, form mathematical categories.

Category theory is a formalism that allows a unified way for expressing properties and constructions that are similar for various structures. Universal algebra is a related subject that studies types of algebraic structures as single objects. For example, the structure of groups is a single object in universal algebra, which is called variety of groups.

BASIC ALGEBRAIC STRUCTURES

Fields

In itself a set is not very useful, being little more than a well-defined collection of mathematical objects. However, when a set has one or more operations (such as addition and multiplication) defined for its elements, it becomes very useful. If the operations satisfy familiar arithmetic rules (such as associativity, commutativity, and distributivity) the set will have a particularly "rich" algebraic structure. Sets with the richest algebraic structure are known as fields.

Familiar examples of fields are the rational numbers (fractions a/b where a and b are positive or negative whole numbers), the real numbers (rational and irrational numbers), and the complex numbers (numbers of the form a + bi where a and b are real numbers and i2 = -1). Each of these is important enough to warrant its own special symbol: Q for the rationals, R for the reals, and C for the complex numbers. The term field in its algebraic sense is quite different from its use in other contexts, such as vector fields in mathematics or magnetic fields in physics. Other languages avoid this conflict in terminology; for example, a field in the algebraic sense is called a corps in French and a Körper in German, both words meaning "body."

In addition to the fields mentioned above, which all have infinitely many elements, there exist fields having only a finite number of elements (always some power of a prime number), and these are of great importance, particularly for discrete mathematics. In fact, finite fields motivated the early development of abstract algebra. The simplest finite field has only two elements, 0 and 1, where 1 + 1 = 0. This field has applications to coding theory and data communication.

Structural axioms

The basic rules, or axioms, for addition and multiplication are shown in the table, and a set that satisfies all 10 of these rules is called a field. A set satisfying only axioms 1–7 is called a ring, and if it also satisfies axiom 9 it is called a ring with unity. A ring satisfying the commutative law of multiplication (axiom 8) is known as a commutative ring.

When axioms 1–9 hold and there are no proper divisors of zero (i.e., whenever ab = 0 either a = 0 or b = 0), a set is called an integral domain. For example, the set of integers {..., -2, -1, 0, 1, 2, ...} is a commutative ring with unity, but it is not a field, because axiom 10 fails. When only axiom 8 fails, a set is known as a division ring or skew field.

Field axioms

axiom 1 Closure: the combination (hereafter indicated by addition or multiplication) of any two elements in the set produces an element in the set.

axiom 2 Addition is commutative: a + b = b + a for any elements in the set.

axiom 3 Addition is associative: a + (b + c) = (a + b) + c for any elements in the set.

axiom 4 Additive identity: there exists an element 0 such that a + 0 = a for every element in the set.

axiom 5 Additive inverse: for each element a in the set, there exists an element -a such that a + (-a) = 0.

axiom 6 Multiplication is associative: a(bc) = (ab)c for any elements in the set.

axiom 7 Distributive law: a(b + c) = ab + ac and (a + b)c = ac + bc for any elements in the set.

axiom 8 Multiplication is commutative: ab = ba for any elements in the set.

axiom 9 Multiplicative identity: there exists an element 1 such that 1a = a for any element in the set.

axiom 10 Multiplicative inverse: for each element a in the set, there exists an element a-1 such that aa-1 = 1.

Quaternions and Abstraction

The discovery of rings having noncommutative multiplication was an important stimulus in the development of modern algebra. For example, the set of n-by-n matrices is a noncommutative ring, but since there are nonzero matrices without inverses, it is not a division ring. The first example of a noncommutative division ring was the quaternions. These are numbers of the form a + bi + cj + dk, where a, b, c, and d are real numbers and their coefficients 1, i, j, and k are unit vectors that define a four-dimensional space.

Quaternions were invented in 1843 by the Irish mathematician William Rowan Hamilton to extend complex numbers from the two-dimensional plane to three dimensions in order to describe physical processes mathematically. Hamilton defined the following rules for quaternion multiplication: i2 = j2 = k2 = -1, ij = k = -ji, jk = i = -kj, and ki = j = -ik. After struggling for some years to discover consistent rules for working with his higher-dimensional complex numbers, inspiration struck while he was strolling in his hometown of Dublin, and he stopped to inscribe these formulas on a nearby bridge. In working with his quaternions, Hamilton laid the foundations for the algebra of matrices and led the way to more abstract notions of numbers and operations.

The Origins of Abstract Algebra

The Axiomatic Approach

The birth of Algebra as mathematicians know it today is also the birth of the axiomatic approach to mathematics.

We generally don't teach students how revolutionary the axiomatic approach is. Typically, in an undergraduate course in Modern Algebra (what is often referred to as a Herstein level course), we simply take the axiomatic approach for granted from Day One. This approach is so familiar and so comfortable to a contemporary mathematician that we seldom give much thought about how bewildering it is to students whose previous experience of mathematics has been limited to courses like calculus. (However sometimes students have seen a little of the axiomatic approach in Linear Algebra, but with very little explanation about why it was ever decided to deal with a seemingly concrete topic like vector spaces in such an abstract way.)

The axiomatic approach is not simply a matter of using axioms in mathematics. The use of axioms, after all, goes back as far as Euclid.

But Euclid, before giving his axioms, starts out by defining the primitive notions of geometry. A point is defined as, roughly speaking, "That which has position but no size." And a line is defined as, "That which has length but no breadth." (I don't remember how Euclid goes about defining the concept of straightness. I don't think he defined a straight line as being the shortest distance between two points.)

The contemporary axiomatic approach, on the other hand, is basically the attitude that when we do mathematics, we don't need to know what the things we are working with are. We only need to know what the rules are. (So that in geometry, we don't need to know what a point or a line is. We only need to know the axioms.)

This is very different from the teaching of mathematics in grammar school and high school and college calculus courses. There, it is considered very important that students understand what numbers are (albeit in a way that to mathematicians is shockingly informal) and what addition, subtraction, and multiplication are, before learning the rules that enable one to actually do arithmetic. And it is very important to thoroughly master arithmetic before going on to represent it in symbolic form in high school algebra. And it is very important to be familiar with a number of specific examples of functions and to understand the concepts of differentiation and integration before going on to learn the rules which enable one to actually differentiate and integrate functions. (In fact, calculus teachers are often annoyed when students, inventing the axiomatic approach on their own, as it were, discover that it is not really necessary to understand the concepts in order to do the calculations.)

But a typical undergraduate course in Modern Algebra starts out saying something like, "A group consists of a set of elements which can be multiplied in such as a way that the following three axioms are satisfied." (Four axioms if one includes closure, which was in fact the key axiom and to some extent was the only one in the original development of group theory, associativity and the existence of an identity element and inverses being taken as pretty much self-evident.)

It is understandable that a student might ask in bewilderment, "But what are these elements? And how does this multiplication work?" And the answer given by the Axiomatic Approach is, "It doesn't matter. Only the rules matter."

This attitude, that it is possible to study things without knowing what one is talking about, is an incredible cognitive leap, and it is the real foundation (not the mathematical foundation, but the psychological foundation) of abstract mathematics.

Of course in order to provide students with the sense that there is some tangible reality to what we are talking about, we immediately provide them with some familiar examples of groups (or rings, or whatever). And one strategy students might use when they can't cope with the level of abstraction they are given is to say, "Okay, when the professor says 'group,' I'm going to think he's talking about the integers. And when he says, 'multiplication,' I'm going to think about addition." (I've used this strategy myself sometimes, when learning a new kind of mathematics.) But this strategy doesn't work very well. It's misleading, because any particular example will have a number of special properties that will not be true of groups in general. (Addition of integers is commutative, for example, and the integers form a cyclic group.) And so if one needs to have concrete things to think about (and I think that almost all of us do), one needs to think not just in terms of one example, but in terms of a number of very dissimilar examples.

And my own experience was that even after I got to be good at this sort of abstract thinking, I would still come to certain concepts (such as the concept of the free product of non-abelian groups, or that of the tensor product) that were so abstract and where it was so difficult to find any natural examples, that for quite a while I still found them very difficult to think about.

But here the axiomatic approach can rescue you on a higher level. You don't really need to think about what a free product or a tensor product actually is ("what it looks like," in my words). You simply need to find a set of axioms that describe the way it behaves. (This has become pretty much the standard way of thinking about tensor products, and I always felt a sort of contempt for mathematicians who proved theorems about tensor products by starting with the construction.)

This is certainly one advantage of the axiomatic approach: that one can work with quite complicated objects (and most mathematical constructions, even the natural numbers, are actually quite complicated) without needing to think about what they "look like." But the primary advantage of the approach is that usually a single set of axioms will describe a very large number of vastly dissimilar mathematical systems, and so by starting from axioms, one can prove theorems that apply to a huge number of different things. (Most of us don't make a big point of using the axiomatic characterization of the real numbers, for instance, because the field of real numbers is the only mathematical thing to which that complete set of axioms applies.)

But how did this revolutionary new way of mathematical thinking come about?

It came about, actually, in a very gradual and somewhat natural way. It came about because over the course of the 19th century, mathematicians started becoming more and more interested in a new kind of subject matter having to do with algebra, but not algebra in the sense of solving equations (although the interest in solving algebraic equations was certainly one of the roots of this new interest). But rather this was algebra in more or less the sense we use the word today (but without thinking of it in abstract terms), namely the study of structures in which one could work in very much the same way that traditional algebra operates in the realm of rational numbers, real numbers, or complex numbers. Some of these structures were: the complex numbers, the quaternions, various algebraic number rings (certain subrings of the complex numbers), in addition to the algebra of matrices developed by Sylvester and Cayley and the algebra of logic developed by Boole. In addition, there was the study of permutation groups, which was originally not thought of as being algebra at all, I believe, but where the basic concepts were developed by Legendre, Abel, and Galois as an approach to understanding the solution of algebraic equations.

All of these subjects were originally studied for very natural and practical reasons having to do with questions in geometry, analysis, number theory, and the theory of equations.

What was new about all these subjects was the interest primarily in the structure as a whole, rather than in doing calculations within that structure. This was perhaps especially clear in the work of Legendre, Abel, and Galois on permutation groups, where what was important was the set of subgroups rather than the individual permutations.

Bourbaki identifies three main streams leading to the development of modern Algebra: (1) The theory of algebraic numbers, developed by Gauss, Dedekind, Kronecker, and Hilbert. (2) The theory of groups of permutations (and, later, groups of geometric transformations), where the work of Galois and Abel was fundamental. (3) The development of linear algebra and hypercomplex systems.

Group Theory

The study of permutation groups was used by Lagrange in the first attempt to develop what would later become Galois Theory. It was Galois who in 1832 later defined the concept of a normal subgroup, a solvable group, and asserted the existence of Sylow subgroups, although apparently without proof. (On the other hand, Galois did not use the concept of a field.)

But as group theory was further developed by other mathematicians (Galois himself, of course, was killed in a duel, apparently because of his political activism, immediately after finishing his treatise), gradually it started becoming clear that the study of permutation groups actually had very little to do with permutations themselves.

And Jordan in 1868 began the study of infinite groups, specifically groups consisting of transformations of geometric space. This study was continued by Felix Klein and Poicaré, and was especially encouraged by Felix Klein's Erlanger Program for geometry. (At this point, there were a number of different kinds of geometry, such as Euclidean geometry, non-Euclidean geometry, projective geometry, affine geometry, and differential geometry. Klein suggested that each particular form of geometry should be characterized as the study of those properties which are invariant under a particular group of transformations. For instance, Euclidean geometry consists of the study of those geometric properties which are not changed by rigid motions.) The concepts and theorems which had been developed for permutation groups applied just as well to these groups of transformations.

By the late 19th century, Cayley and Dedekind and many other mathematicians were becoming very aware that what was really relevant in group theory was the law of composition (multiplication) in a group and not the nature of the objects making up the group.

But the importance of groups at this point still had to do with their concrete applications. Groups were still seen as consisting of operators of some sort and Dedekind and Cayley stopped short of defining groups in an axiomatic way and seeing them as structures which were of interest for their own sake.

Algebraic Number Theory and the Rise of Commutative Ring Theory

The theory of algebraic numbers (which was initiated by Gauss) was primarily developed in connection with attempts to prove Fermat's Last Theorem. Lagrange, Euler, and Gauss had seen that an approach to proving this is to notice that the left-hand side of the equation $xp + yp = zp$ factors completely if one allows the use of the pth root of -1, which is of course a complex number. (It's adequate to deal with the case where p is prime and of course different from 2.) It is commonly believed that this is the proof that Fermat himself had in mind. But the proof is flawed, because it depends on the assumption that factorization into prime factors is unique, even when algebraic numbers are involved. But this is not always true, as one can see from the example,

$$\left(1+\sqrt{-5}\right)\left(1+\sqrt{-5}\right)=(2)(3),$$

where it is easy to prove that none of four factors in these two decompositions can be further decomposed into products within the ring consisting of all numbers of the form a + b $\sqrt{-5}$, with a and b ordinary integers. (Here $\sqrt{-5}$ is used to denote the square root of -5, which is, of course, a complex number.)

The theory of algebraic numbers was further developed by Dirichlet, Hermite, Kummer, Kronecker, and Dedekind. Kronecker and Dedekind used two different methods (which although very dissimilar are ultimately equivalent) to introduce of certain "ideal numbers" into algebraic number rings to remedy the lack of unique factorization. Dedekind's method was the invention of what we today call, in an arbitrary ring, ideals. In his work, Dedekind basically established the foundations of modern commutative ring theory. However the methods of Dedekind and Kronecker fell short of providing a proof of Fermat's Last Theorem, although they did enable proofs in many special cases.

The other main thread leading to modern commutative ring theory came from algebraic geometry, and I won't really discuss that here except to mention that mathematicians were becoming very aware that the algebra of functions defined on an algebraic curve or surface had a great deal in common with algebraic number rings. Here we see that importance of the fact that mathematicians working in what originally seemed very different specialties were familiar with each others work and influenced by it. (There existed still a third major example of commutative rings, namely those consisting of functions defined by power series.)

Bourbaki identifies the 142 page article by Steinitz in 1910 titled The Algebraic Theory of Fields as having given birth to the modern concept of Algebra. (One can also note that much earlier, Peano, in 1888, gave the axiomatic definition of a real vector space and defined the concept of a linear transformation between vector spaces.) The word field had first been used by Dedekind, whose concern was with certain fields contained within the complex numbers (algebraic number fields). And it was Dedekind and Hilbert who had first seen Galois Theory as a correspondance between subfields and subgroups of the Galois group. (Dedekind was the first to think of the Galois group as consisting of the automorphisms of the field extension rather than permutations of the roots of the polynomial in question.)

Steinitz in his 1910 article developed the notions of prime field (by this time there had been a lot of work on the theory of finite fields), separable extension, and transcendence degree, and proved that every field has an algebraically closed extension. But what makes his article thoroughly modern is that instead of defining a field as a set of complex numbers or congruence classes or the like, Steinitz simply defined a field to be a structure consisting of a set of elements in which two operations are defined (to be referred to as addition and multiplication) satisfying a certain set of rules.

The concept of a ring was first used by Dedekind, who used the word "order" (or rather, of course, its German equivalent, "ordnung.") The word "ring" (which is actually the same in German and in English) was introduced by Hilbert. The point is that in an algebraic number ring (or any finite integral extension of a base ring), if one looks at the powers of an element then one finds a point where subsequent powers can be expressed as linear combinations of the preceding ones. Thus the multiplication in a certain sense turns back on itself in a way that is somewhat like a geometric ring.

But it was not until 1914 when the first paper where the general notion of a ring is defined axiomatically: "On Zero Divisors and Decomposition of Rings," by Fraenkel. Although this gave the general definition, the paper itself was concerned with commutative artinian local rings where the unique prime ideal is principal.

In the same year, 1914, Hausdorff in his Grundzüge der Mengenlehre, gave an axiomatic definition of general topology. Of course this was a time when algebraists, analysts, and topologists talked to each other, were interested in each other's work, were influenced by each other, and in many cases were actually the same individuals.

The Theory of Algebras (Hypercomplex Systems)

The starting point for third stream in the development of algebra was Hamilton's invention of the quaternions. (Complex numbers had been used in algebra by the Italian algebraists since roughly 1550 and by 1800 the theory of calculus using complex variables was fairly well developed, by Cauchy and others. It was Gauss who introduced the geometric representation of complex numbers.)

In 1878, Frobenius proved that the quaternions were the only possible (finite-dimensional) associative extension of the complex numbers in which division was possible and the only (finite-dimensional) non-commutative extension of the real numbers in which division was possible. This was independently proved two years later by C. S. Pierce. (Gauss had been convinced that the field of complex numbers was the only finite dimensional commutative field extension of the real number system. This was subsequently proved by Weierstrass.)

Later Cayley noted that there exists a set of two by two matrices satisfyiing the multiplication table of the quaternions. (The concept of a matrix is due to Sylvester, who introduced matrices as a shorthand for substitutions of variables, i.e. what we now call linear transformations.) But not until about 1870 was it noted, by the Americans B. Pierce and C.S. Pierce, that the set of square matrices of a given size form an algebraic system which permits addition, subtraction, and multiplication (i.e., in modern terminology, a ring).

The term "an algebra" seems to have been used by the Americans and British in pretty much its modern sense, i.e. a ring which is a finite-dimensional vector space over the complex numbers (or real numbers). On the other hand, the Germans generally preferred the term "hypercomplex system." Aside from Hamilton's quaternions, the main example before 1850 was Grassman's "exterior algebras," but the analogy to the quaternions and other algebras was only much later seen.

Other examples of algebras over the complex numbers were seen during the period 1850 to 1860, but the general study of algebras (and thus the roots of non-commutative ring theory) begins only in 1870 in the work of B. Pierce and C.S. Pierce, who introduce the concepts of idempotent and nilpotent elements and the decomposition of an idempotent element into a sum of orthogonal primitive idempotents.

Cayley and Sylvester and other British and American mathematicians then started working on the problem of classifying algebras of small dimension over the complex numbers.

During this time, the development of Lie groups and algebras (which are non-associative) was proceeding and some of the fundamental concepts in the theory of associative algebras (the concept of the radical, for instance) were developed first for Lie algebras.

Another key source of ideas and examples was the concept of a group algebra, which had been essentially defined by Dedekind in 1896, in a letter to Frobenius. Dedekind was very clear on the relation of this to the general theory of algebras, although the theory of group representations as developed by Burnside and Schur (around 1905) did not at that time explicitly use ring-theoretic methods.

The concept of a simple algebra over the complex numbers had been defined in 1893 by the German mathematician T. Molien, who then proved the first version of the Wedderburn Theorem, i.e. that a simple algebra over the complex numbers is isomorphic to the ring of n by n matrices over the complex numbers. It was at this point that the concept of a two-sided ideal became current and a lot of theorems were proved about them. But it is not clear from the Bourbaki survey whether the word "ideal" was originally used, and it is possible that the analogy to Dedekind's work on commutative rings was not immediately seen.

The concept of a semi-simple algebra was introduced by Elie Cartan. (Unfortunately, I don't have a date.)

The development around 1900 of the theory of finite fields by the American mathematicians E. H. Moore and L. E. Dickson was what motivated the generalization of the theory of algebras to the case where the base field was unrestricted. Wedderburn, another American, in 1905 proved that every finite skew field (a.k.a. division algebra) is in fact commutative.

In 1903, in a memoir on the algebraic solution of differential equations, Poincaré had defined the concepts of left ideal and right ideal for an algebra. (As mentioned, two-sided ideals had been essentially known since Molien's paper in 1893.) In this memoir, Poincaré proves that the minimal left ideals in the ring of n by n matrices have dimension n. However this result was not noticed by the algebraists.

The notions of left and right ideals were rediscovered in 1907 by Wedderburn, who proved that the radical was the largest nilpotent left ideal and proved his well known "Wedderburn Theorem" (later generalized by Emil Artin) which states that every semi-simple algebra over an arbitrary base field is a direct product of matrix rings over skew fields.

In 1920, Emmy Nöther and W. Schmeidler used the concepts of left and right ideal in a paper devoted to rings of differential operators. But otherwise, these concepts were ignored after Wedderburn's paper until 1927, when Emmy Nöther, and Brauer (and later A.A. Albert and Hasse) resumed the study of them.

By 1934, the basic theory of semi-simple rings was essentially complete.

Bourbaki's summary statement is, "The axiomatization of algebra was begun by Dedekind and and Hilbert, and then vigorously pursued by Steinitz (1910). It was then completed in the years following 1920 by Artin, Nöther and their colleagues at Göttingen (Hasse, Krull, Schreier, van der Waerden). It was presented to the world in complete form by van der Waerden's book (1930)."

What we see from all this (at least in my view) is that the development of modern Algebra was never motivated by mathematicians seeking abstraction for its own sake. Instead, algebraists working on quite concrete problems were trying to invent tools that might help with their investigation of these problems, and slowly (very slowly, as we look at their work in retrospect) began to notice that the same logical patterns recurred over and over again in different examples.

Brief History of Modern Algebra

As in other parts of mathematics, concrete problems and examples have played important roles in the development of abstract algebra.Through the end of the nineteenth century, many – perhaps most – of these problems were in some way related to the theory of algebraic equations. Major themes include:

- Solving of systems of linear equations, which led to linear algebra
- Attempts to find formulas for solutions of general polynomial equations of higher degree that resulted in discovery of groups as abstract manifestations of symmetry
- Arithmetical investigations of quadratic and higher degree forms and diophantine equations, that directly produced the notions of a ring and ideal.

Numerous textbooks in abstract algebra start with axiomatic definitions of various algebraic structures and then proceed to establish their properties. This creates a false impression that in algebra axioms had come first and then served as a motivation and as a basis of further study. The true order of historical development was almost exactly the opposite. For example, the hypercomplex numbers of the nineteenth century had kinematic and physical motivations but challenged comprehension. Most theories that are now recognized as parts of algebra started as collections of disparate facts from various branches of mathematics, acquired a common theme that served as a core around which various results were grouped, and finally became unified on a basis of a common set of concepts. An archetypical example of this progressive synthesis can be seen in the history of group theory.

Early Group Theory

There were several threads in the early development of group theory, in modern language loosely corresponding to number theory, theory of equations, and geometry.Leonhard Euler considered algebraic operations on numbers modulo an integer, modular arithmetic, in his generalization of Fermat's little theorem. These investigations were taken much further by Carl Friedrich Gauss, who considered the structure of multiplicative groups of residues mod n and established many properties of cyclic and more general abelian groups that arise in this way. In his investigations of composition of binary quadratic forms, Gauss explicitly stated the associative law for the composition of forms, but like Euler before him, he seems to have been more interested in concrete results than in general theory.

In 1870, Leopold Kronecker gave a definition of an abelian group in the context of ideal class groups of a number field, generalizing Gauss's work; but it appears he did not tie his definition with previous work on groups, particularly permutation groups. In 1882, considering the same question, Heinrich M. Weber realized the connection and gave a similar definition that involved the cancellation property but omitted the existence of the inverse element, which was sufficient in his context (finite groups).

Permutations were studied by Joseph-Louis Lagrange in his 1770 paper Réflexions sur la résolution algébrique des équations (Thoughts on the algebraic solution of equations) devoted to solutions of algebraic equations, in which he introduced Lagrange resolvents. Lagrange's goal was to understand why equations of third and fourth degree admit formulas for solutions, and he identified as key objects permutations of the roots. An important novel step taken by Lagrange in this paper was the abstract view of the roots, i.e. as symbols and not as numbers. However, he did not consider composition of permutations. Serendipitously, the first edition of Edward Waring's Meditationes Algebraicae (Meditations on Algebra) appeared in the same year, with an expanded version published in 1782. Waring proved the fundamental theorem of symmetric polynomials, and specially considered the relation between the roots of a quartic equation and its resolvent cubic. Mémoire sur la résolution des équations (Memoire on the Solving of Equations) of Alexandre Vandermonde (1771) developed the theory of symmetric functions from a slightly different angle, but like Lagrange, with the goal of understanding solvability of algebraic equations.

Kronecker claimed in 1888 that the study of modern algebra began with this first paper of Vandermonde. Cauchy states quite clearly that Vandermonde had priority over Lagrange for this remarkable idea, which eventually led to the study of group theory.

Paolo Ruffini was the first person to develop the theory of permutation groups, and like his predecessors, also in the context of solving algebraic equations. His goal was to establish the impossibility of an algebraic solution to a general algebraic equation of degree greater than four. En route to this goal he introduced the notion of the order of an element of a group, conjugacy, the cycle decomposition of elements of permutation groups and the notions of primitive and imprimitive and proved some important theorems relating these concepts, such as if G is a subgroup of S5 whose order is divisible by 5 then G contains an element of order 5. Note, however, that he got by without formalizing the concept of a group, or even of a permutation group. The next step was taken by Évariste Galois in 1832, although his work remained unpublished until 1846, when he considered for the first time what is now called the closure property of a group of permutations, which he expressed as if in such a group one has the substitutions S and T then one has the substitution ST.

The theory of permutation groups received further far-reaching development in the hands of Augustin Cauchy and Camille Jordan, both through introduction of new concepts and, primarily, a great wealth of results about special classes of permutation groups and even some general theorems. Among other things, Jordan defined a notion of isomorphism, still in the context of permutation groups and, incidentally, it was he who put the term group in wide use.

The abstract notion of a group appeared for the first time in Arthur Cayley's papers in 1854. Cayley realized that a group need not be a permutation group (or even finite), and may instead consist of matrices, whose algebraic properties, such as multiplication and inverses, he systematically investigated in succeeding years. Much later Cayley would revisit the question whether abstract groups were more general than permutation groups, and establish that, in fact, any group is isomorphic to a group of permutations.

Modern Algebra

The end of the 19th and the beginning of the 20th century saw a tremendous shift in the methodology of mathematics. Abstract algebra emerged around the start of the 20th century, under the name modern algebra. Its study was part of the drive for more intellectual rigor in mathematics. Initially, the assumptions in classical algebra, on which the whole of mathematics (and major parts of the natural sciences) depend, took the form of axiomatic systems. No longer satisfied with establishing properties of concrete objects, mathematicians started to turn their attention to general theory. Formal definitions of certain algebraic structures began to emerge in the 19th century. For example, results about various groups of permutations came to be seen as instances of general theorems that concern a general notion of an abstract group. Questions of structure and classification of various mathematical objects came to forefront.

These processes were occurring throughout all of mathematics, but became especially pronounced in algebra. Formal definition through primitive operations and axioms were proposed for many basic algebraic structures, such as groups, rings, and fields. Hence such things as group theory and ring theory took their places in pure mathematics. The algebraic investigations of general fields by Ernst Steinitz and of commutative and then general rings by David Hilbert, Emil Artin and Emmy Noether, building up on the work of Ernst Kummer, Leopold Kronecker and Richard Dedekind, who had considered ideals in commutative rings, and of Georg Frobenius and Issai Schur, concerning representation theory of groups, came to define abstract algebra. These developments of the last quarter of the 19th century and the first quarter of 20th century were systematically exposed in Bartel van der Waerden's Moderne algebra, the two-volume monograph published in 1930–1931 that forever changed for the mathematical world the meaning of the word algebra from the theory of equations to the theory of algebraic structures.

Basic Concepts

By abstracting away various amounts of detail, mathematicians have defined various algebraic structures that are used in many areas of mathematics. For instance, almost all systems studied are sets, to which the theorems of set theory apply. Those sets that have a certain binary operation defined on them form magmas, to which the concepts concerning magmas, as well those concerning sets, apply. We can add additional constraints on the algebraic structure, such as associativity (to form semigroups); identity, and inverses (to form groups); and other more complex structures. With additional structure, more theorems could be proved, but the generality is reduced. The "hierarchy" of algebraic objects (in terms of generality) creates a hierarchy of the corresponding theories: for instance, the theorems of group theory may be used when studying rings (algebraic objects that have two binary operations with certain axioms) since a ring is a group over one of its operations. In general there is a balance between the amount of generality and the richness of the theory: more general structures have usually fewer nontrivial theorems and fewer applications.

Examples of algebraic structures with a single binary operation are:

- Magma
- Quasigroup
- Monoid
- Semigroup
- Group

Examples involving several operations include:

- Ring
- Field
- Module
- Vector space
- Algebra over a field
- Associative algebra
- Lie algebra
- Lattice
- Boolean algebra
- Algebraic structure

In mathematics, and more specifically in abstract algebra, an algebraic structure on a set A is a collection of finitary operations on A; the set A with this structure is also called an algebra. Examples of algebraic structures include groups, rings, fields, and lattices. More complex structures can be defined by introducing multiple operations, different underlying sets, or by altering the defining axioms. Examples of more complex algebraic structures include vector spaces, modules, and algebras.

The properties of specific algebraic structures are studied in abstract algebra. The general theory of algebraic structures has been formalized in universal algebra. The language of category theory is used to express and study relationships between different classes of algebraic and non-algebraic objects. This is because it is sometimes possible to find strong connections between some classes of objects, sometimes of different kinds. For example, Galois theory establishes a connection between certain fields and groups: two algebraic structures of different kinds.

Addition and multiplication on numbers are the prototypical example of an operation that combines two elements of a set to produce a third. These operations obey several algebraic laws. For example, a + (b + c) = (a + b) + c and a(bc) = (ab)c, both examples of the associative law. Also a + b = b + a, and ab = ba, the commutative law. Many systems studied by mathematicians have operations that obey some, but not necessarily all, of the laws of ordinary arithmetic. For example, rotations of objects in three-dimensional space can be combined by performing the first rotation and then applying the second rotation to the object in its new orientation. This operation on rotations obeys the associative law, but can fail the commutative law.

Mathematicians give names to sets with one or more operations that obey a particular collection of laws, and study them in the abstract as algebraic structures. When a new problem can be shown to follow the laws of one of these algebraic structures, all the work that has been done on that category in the past can be applied to the new problem. In full generality, algebraic structures may involve an arbitrary number of sets and operations that can combine more than two elements (higher arity), but this article focuses on binary operations on one or two sets. The examples here are by no means a complete list, but they are meant to be a representative list and include the most common structures. Longer lists of algebraic structures may be found in the external links and within Category:Algebraic structures. Structures are listed in approximate order of increasing complexity.

EXAMPLES

One set With Operations

Simple structures: no binary operation:

- Set: a degenerate algebraic structure S having no operations.
- Pointed set: S has one or more distinguished elements, often 0, 1, or both.
- Unary system: S and a single unary operation over S.
- Pointed unary system: a unary system with S a pointed set.

Group-like structures: one binary operation. The binary operation can be indicated by any symbol, or with no symbol (juxtaposition) as is done for ordinary multiplication of real numbers.

- Magma or groupoid: S and a single binary operation over S.
- Semigroup: an associative magma.
- Monoid: a semigroup with Identity element.
- Group: a monoid with a unary operation (inverse), giving rise to inverse elements.
- Abelian group: a group whose binary operation is commutative.
- Semilattice: a semigroup whose operation is idempotent and commutative. The binary operation can be called either meet or join.
- Quasigroup: a magma obeying the latin square property. A quasigroup may also be represented using three binary operations.
- Loop: a quasigroup with identity.

Ring-like structures or Ringoids: two binary operations, often called addition and multiplication, with multiplication distributing over addition.

- Semiring: a ringoid such that S is a monoid under each operation. Addition is typically assumed to be commutative and associative, and the monoid product is assumed to distribute over the addition on both sides, and the additive identity satisfies 0/ x = 0 for all x.
- Near-ring: a semiring whose additive monoid is a (not necessarily abelian) group.
- Ring: a semiring whose additive monoid is an abelian group.
- Lie ring: a ringoid whose additive monoid is an abelian group, but whose multiplicative operation satisfies the Jacobi identity rather than associativity.
- Boolean ring:a commutative ring with idempotent multiplication operation.
- Field: a commutative ring which contains a multiplicative inverse for every nonzero element
- Kleene algebras: a semiring with idempotent addition and a unary operation, the Kleene star, satisfying additional properties.
- *-algebra: a ring with an additional unary operation (*) satisfying additional properties.

Lattice structures: two or more binary operations, including operations called meet and join, connected by the absorption law.

- Complete lattice: a lattice in which arbitrary meet and joins exist.

- Bounded lattice: a lattice with a greatest element and least element.
- Complemented lattice: a bounded lattice with a unary operation, complementation, denoted by postfix ¥". The join of an element with its complement is the greatest element, and the meet of the two elements is the least element.
- Modular lattice: a lattice whose elements satisfy the additional modular identity.
- Distributive lattice: a lattice in which each of meet and join distributes over the other. Distributive lattices are modular, but the converse does not hold.
- Boolean algebra: a complemented distributive lattice. Either of meet or join can be defined in terms of the other and complementation. This can be shown to be equivalent with the ring-like structure of the same name above.
- Heyting algebra: a bounded distributive lattice with an added binary operation, relative pseudo-complement, denoted by infix '!, and governed by the axiomsx/ '!/ x = 1,x (x-y) = xy,y (x -y) = y,x- (y-z) = (x-y) (x-z).

Arithmetics: two binary operations, addition and multiplication. S is an infinite set. Arithmetics are pointed unary systems, whose unary operation is injective successor, and with distinguished element 0.

- Robinson arithmetic. Addition and multiplication are recursively defined by means of successor. 0 is the identity element for addition, and annihilates multiplication. Robinson arithmetic is listed here even though it is a variety, because of its closeness to Peano arithmetic.

Peano arithmetic. Robinson arithmetic with an axiom schema of induction. Most ring and field axioms bearing on the properties of addition and multiplication are theorems of Peano arithmetic or of proper extensions thereof.

Two sets with Operations

Module-like structures: composite systems involving two sets and employing at least two binary operations.

- Group with operators: a group G with a set ? and a binary operation ×GG satisfying certain axioms.
- Module: an abelian group M and a ring R acting as operators on M. The members of R are sometimes called scalars, and the binary operation of scalar multiplication is a function R×M- M, which satisfies several axioms. Counting the ring operations these systems have at least three operations.

- Vector space: a module where the ring R is a division ring or field.
- Graded vector space: a vector space with a direct sum decomposition breaking the space into "grades".

Quadratic space: a vector space V over a field F with a function from V into F satisfying certain properties. Every quadratic space is also an inner product space (see below). Algebra-like structures: composite system defined over two sets, a ring R and an R-module M equipped with an operation called multiplication. This can be viewed as a system with five binary operations: two operations on R, two on M and one involving both R and M.

- Algebra over a ring (also R-algebra): a module over a commutative ring R, which also carries a multiplication operation that is compatible with the module structure. This includes distributivity over addition and linearity with respect to multiplication by elements of R. The theory of an algebra over a field is especially well developed.
- Associative algebra: an algebra over a ring such that the multiplication is associative.
- Nonassociative algebra: a module over a commutative ring, equipped with a ring multiplication operation that is not necessarily associative. Often associativity is replaced with a different identity, such as alternation, the Jacobi identity, or the Jordan identity.
- Coalgebra: a vector space with a "comultiplication" defined dually to that of associative algebras.
- Lie algebra: a special type of nonassociative algebra whose product satisfies the Jacobi identity.•
- Lie coalgebra: a vector space with a "comultiplication" defined dually to that of Lie algebras.
- Graded algebra: a graded vector space with an algebra structure compatible with the grading. The idea is that if the grades of two elements a and b are known, then the grade of ab is known, and so the location of the product ab is determined in the decomposition.

Inner product space: an F vector space V with a sesquilinear binary operation V × VF.

Four or more binary operations:

- Bialgebra: an associative algebra with a compatible coalgebra structure.

- Lie bialgebra: a Lie algebra with a compatible bialgebra structure.
- Hopf algebra: a bialgebra with a connection axiom (antipode).
- Clifford algebra: a graded associative algebra equipped with an exterior product from which may be derived several possible inner products. Exterior algebras and geometric algebras are special cases of this construction.

HYBRID STRUCTURES

Algebraic structures can also coexist with added structure of non-algebraic nature, such as partial order or a topology. The added structure must be compatible, in some sense, with the algebraic structure.

- Topological group: a group with a topology compatible with the group operation.
- Lie group: a topological group with a compatible smooth manifold structure.Ordered groups, ordered rings and ordered fields: each type of structure with a compatible partial order.
- Archimedean group: a linearly ordered group for which the Archimedean property holds.
- Topological vector space: a vector space whose M has a compatible topology.
- Normed vector space: a vector space with a compatible norm. If such a space is complete (as a metric space) then it is called a Banach space.
- Hilbert space: an inner product space over the real or complex numbers whose inner product gives rise to a Banach space structure.
- Vertex Operator Algebra
- Von Neumann algebra: a *-algebra of operators on a Hilbert space equipped with the weak operator topology.

Binary Operation

A binary operation is a calculation that combines the arguments x and y to In mathem atics, a binary operation or dyadic operation is a calculation that combines two elements to produce another element. More formally, a binary operation is an operation of arity two. More specifically, a binary operation on a set is a binary operation whose two domains and the codomain are the same set. Examples include the familiar arithmetic operations of addition, subtraction, multiplication. Other examples are readily found in different areas of mathematics, such as vector addition, matrix multiplication and conjugation in groups.

However, a binary operation may also involve several sets. For example, scalar multiplication of vector spaces takes a scalar and a vector to produce a vector, and scalar product takes two vectors to produce a scalar. Binary operations are the keystone of most algebraic structures, that are studied in algebra, and used in all mathematics, such as fields, groups, monoids, rings, algebras, and many more.

Terminology

More precisely, a binary operation on a set S is a map which sends elements of the Cartesian product S × S to S: Because the result of performing the operation on a pair of elements of S is again an element of S, the operation is called a closed binary operation on S (or sometimes expressed as having the property of closure). If f is not a function, but is instead a partial function, it is called a partial binary operation. For instance, division of real numbers is a partial binary operation, because one can't divide by zero: a/0 is not defined for any real a. However, both in universal algebra and model theory the binary operations considered are defined on all of S × S. Sometimes, especially in computer science, the term is used for any binary function.

Properties and Examples

Typical examples of binary operations are the addition (+) and multiplication (×) of numbers and matrices as well as composition of functions on a single set. For instance,

- On the set of real numbers R, f(a, b) = a + b is a binary operation since the sum of two real numbers is a real number.
- On the set of natural numbers N, f(a, b) = a + b is a binary operation since the sum of two natural numbers is a natural number. This is a different binary operation than the previous one since the sets are different.
- On the set M(2,R) of 2 × 2 matrices with real entries, f(A, B) = A + B is a binary operation since the sum of two such matrices is a 2 × 2 matrix.
- On the set M(2,R) of 2 × 2 matrices with real entries, f(A, B) = AB is a binary operation since the product of two such matrices is a 2 × 2 matrix.
- For a given set C, let S be the set of all functions h : C '! C. Define f : S × S '! S by f(h1, h2)(c) = (h1 "' h2) (c) = h1(h2(c)) for all c " C, the composition of the two functions h1 and h2 in S. Then f is a binary operation since the composition of the two functions is again a function on the set C (that is, a member of S).

Many binary operations of interest in both algebra and formal logic are commutative, satisfying f(a, b) = f(b, a) for all elements a and b in S, or associative, satisfying f(f(a, b), c) = f(a, f(b, c)) for all a, b and c in S. Many also have identity elements and inverse elements.

The first three examples above are commutative and all of the above examples are associative.On the set of real numbers R, subtraction, that is, f(a, b) = a " b, is a binary operation which is not commutative since, in general, a " b `" b " a. It is also not associative, since, in general, a " (b " c) `" (a " b) " c; for instance, 1 " (2 " 3) = 2 but (1 " 2) " 3 = "4. On the set of natural numbers N, the binary operation exponentiation, f(a,b) = ab, is not commutative since, ab "" ba (cf. Equation $x_{,}$ = yã), and is also not associative since f(f(a, b), c) "" f(a, f(b, c)). For instance, with a = 2, b = 3 and c = 2, f(23,2) = f(8,2) = 82 = 64, but f(2,32) = f(2,9) = 29 = 512. By changing the set N to the set of integers Z, this binary operation becomes a partial binary operation since it is now undefined when a = 0 and b is any negative integer. For either set, this operation has a right identity (which is 1) since f(a, 1) = a for all a in the set, which is not an identity (two sided identity) since f(1, b) "" b in general.

Division (/), a partial binary operation on the set of real or rational numbers, is not commutative or associative. Tetration (), as a binary operation on the natural numbers, is not commutative or associative and has no identity element.

Notation

Binary operations are often written using infix notation such as a " b, a + b, a · b or (by juxtaposition with no symbol) ab rather than by functional notation of the form f(a, b). Powers are usually also written without operator, but with the second argument as superscript. Binary operations sometimes use prefix or (probably more often) postfix notation, both of which dispense with parentheses. They are also called, respectively, Polish notation and reverse Polish notation.

Pair and Tuple

A binary operation, ab, depends on the ordered pair (a, b) and so (ab)c (where the parentheses here mean first operate on the ordered pair (a, b) and then operate on the result of that using the ordered pair ((ab), c)) depends in general on the ordered pair ((a, b), c). Thus, for the general, non-associative case, binary operations can be represented with binary trees.

However:

- If the operation is associative, (ab)c = a(bc), then the value of (ab)c depends only on the tuple (a, b, c).
- If the operation is commutative, (ab)c = a(bd), then the value of (ab)c depends only on (a, b, c), where braces indicate multisets.

- If the operation is both associative and commutative then the value of (ab) c depends only on the multiset {a, b, c}.
- If the operation is associative, commutative and idempotent, aa = a, then the value of (ab)c depends only on the set {a, b, c}.

Binary Operations as Ternary Relations

A binary operation f on a set S may be viewed as a ternary relation on S, that is, the set of triples (a, b, f(a,b)) in S × S × S for all a and b in S.

External Binary Operations

An external binary operation is a binary function from K × S to S. This differs from a binary operation on a set in the sense in that K need not be S; its elements come from outside. An example of an external binary operation is scalar multiplication in linear algebra. Here K is a field and S is a vector space over that field. An external binary operation may alternatively be viewed as an action; K is acting on S. The dot product of two vectors maps from S × S to K, where K is a field and S is a vector space over K. It depends on authors whether it is considered as a binary operation.

Universal Algebra

Algebraic structures are defined through different configurations of axioms. Universal algebra abstractly studies such objects. One major dichotomy is between structures that are axiomatized entirely by identities and structures that are not. If all axioms defining a class of algebras are identities, then this class is a variety (not to be confused with algebraic varieties of algebraic geometry). Identities are equations formulated using only the operations the structure allows, and variables that are tacitly universally Universal_quantifier" quantified over the relevant universe. Identities contain no connectives, existentially quantified variables, or relations of any kind other than the allowed operations. The study of varieties is an important part of universal algebra. An algebraic structure in a variety may be understood as the quotient algebra of term algebra (also called "absolutely free algebra") divided by the equivalence relations generated by a set of identities. So, a collection of functions with given signatures generate a free algebra, the term algebra T. Given a set of equational identities (the axioms), one may consider their symmetric, transitive closure E. The quotient algebra T/E is then the algebraic structure or variety. Thus, for example, groups have a signature containing two operators: the multiplication operator m, taking two arguments, and the inverse operator i, taking one argument, and the identity element e, a constant, which may be considered an operator that takes zero arguments. Given a (countable) set of variables x, y, z, etc. the term algebra is the collection of all possible terms involving m, i, e and the variables; so for example, m(i(x), m(x,m(y,e))) would be an element of the term algebra. One of the axioms defining a group is the identity m(x, i(x)) = e; another is m(x,e) = x. The axioms can be represented as trees. These equations induce equivalence classes on the free algebra; the quotient algebra then has the algebraic structure of a group.

Some structures do not form varieties, because either:

- It is necessary that 0 “‘ 1, 0 being the additive identity element and 1 being a multiplicative identity element, but this is a nonidentity;
- Structures such as fields have some axioms that hold only for nonzero members of S. For an algebraic structure to be a variety, its operations must be defined for all members of S; there can be no partial operations.

Structures whose axioms unavoidably include nonidentities are among the most important ones in mathematics, e.g., fields and division rings. Structures with nonidentities present challenges varieties do not. For example, the direct product of two fields is not a field, because (1,0).(0,1)=(0,0), but fields do not have zero divisors.

Category Theory

Category theory is another tool for studying algebraic structures (see, for example, Mac Lane 1998). A category is a collection of objects with associated morphisms. Every algebraic structure has its own notion of homomorphism, namely any function compatible with the operation(s) defining the structure. In this way, every algebraic structure gives rise to a category. For example, the category of groups has all groups as objects and all group homomorphisms as morphisms. This concrete category may be seen as a category of sets with added category-theoretic structure. Likewise, the category of topological groups (whose morphisms are the continuous group homomorphisms) is a category of topological spaces with extra structure. A forgetful functor between categories of algebraic structures "forgets" a part of a structure.

There are various concepts in category theory that try to capture the algebraic character of a context, for instance

- Algebraic category
- Essentially algebraic category
- Presentable category

- Locally presentable category
- Monadic functors and categories

Different Meanings of "Structure"

In a slight abuse of notation, the word "structure" can also refer to just the operations on a structure, instead of the underlying set itself. For example, the sentence, "We have defined a ring structure on the set," means that we have defined ring operations on the set A. For another example, the group $(\mathbf{z},+)$ can be seen as a set $\mathbf{z}$ that is equipped with an algebraic structure, namely the operation.

Applications

Because of its generality, abstract algebra is used in many fields of mathematics and science. For instance, algebraic topology uses algebraic objects to study topologies. The Poincaré conjecture, proved in 2003, asserts that the fundamental group of a manifold, which encodes information about connectedness, can be used to determine whether a manifold is a sphere or not. Algebraic number theory studies various number rings that generalize the set of integers. Using tools of algebraic number theory, Andrew Wiles proved Fermat's Last Theorem.

In physics, groups are used to represent symmetry operations, and the usage of group theory could simplify differential equations. In gauge theory, the requirement of local symmetry can be used to deduce the equations describing a system. The groups that describe those symmetries are Lie groups, and the study of Lie groups and Lie algebras reveals much about the physical system; for instance, the number of force carriers in a theory is equal to the dimension of the Lie algebra, and these bosons interact with the force they mediate if the Lie algebra is nonabelian.

2

THE DEVELOPMENT OF RING THEORY

In mathematics, rings are algebraic structures that generalize fields: multiplication need not be commutative and multiplicative inverses need not exist. In other words, a ring is a set equipped with two binary operations satisfying properties analogous to those of addition and multiplication of integers. Ring elements may be numbers such as integers or complex numbers, but they may also be non-numerical objects such as polynomials, square matrices, functions, and power series.

Formally, a ring is an abelian group whose operation is called addition, with a second binary operation called multiplication that is associative, is distributive over the addition operation, and has a multiplicative identity element. (Some authors use the term "rng" with a missing i to refer to the more general structure that omits this last requirement; see § Notes on the definition.)

Whether a ring is commutative (that is, whether the order in which two elements are multiplied might change the result) has profound implications on its behavior. Commutative algebra, the theory of commutative rings, is a major branch of ring theory. Its development has been greatly influenced by problems and ideas of algebraic number theory and algebraic geometry. The simplest commutative rings are those that admit division by non-zero elements; such rings are called fields.

Examples of commutative rings include the set of integers with their standard addition and multiplication, the set of polynomials with their addition and multiplication, the coordinate ring of an affine algebraic variety, and the ring of integers of a number field. Examples of noncommutative rings include the ring of n × n real square matrices with n = 2, group rings in representation theory, operator algebras in functional analysis, rings of differential operators, and cohomology rings in topology.

The conceptualization of rings spanned the 1870s to the 1920s, with key contributions by Dedekind, Hilbert, Fraenkel, and Noether. Rings were first formalized as a generalization of Dedekind domains that occur in number theory, and of polynomial rings and rings of invariants that occur in algebraic geometry and invariant theory. They later proved useful in other branches of mathematics such as geometry and analysis.

Commutative algebra, a major area of modern mathematics. Because these three fields (algebraic geometry, algebraic number theory and commutative algebra) are so intimately connected it is usually difficult and meaningless to decide which field a particular result belongs to. For example, Hilbert's Nullstellensatz is a theorem which is fundamental for algebraic geometry, and is stated and proved in terms of commutative algebra. Similarly, Fermat's last theorem is stated in terms of elementary arithmetic, which is a part of commutative algebra, but its proof involves deep results of both algebraic number theory and algebraic geometry.

Noncommutative rings are quite different in flavour, since more unusual behavior can arise. While the theory has developed in its own right, a fairly recent trend has sought to parallel the commutative development by building the theory of certain classes of noncommutative rings in a geometric fashion as if they were rings of functions on (non-existent) 'noncommutative spaces'. This trend started in the 1980s with the development of noncommutative geometry and with the discovery of quantum groups. It has led to a better understanding of noncommutative rings, especially noncommutative Noetherian rings. For the definitions of a ring and basic concepts and their properties, see ring (mathematics). The definitions of terms used throughout ring theory may be found in the glossary of ring theory.

RINGS IN NUMBER THEORY

In another direction, important progress in number theory by German mathematicians such as Ernst Kummer, Richard Dedekind, and Leopold Kronecker used rings of algebraic integers. (An algebraic integer is a complex number satisfying an algebraic equation of the form xn + a1xn-1 + … + an = 0 where the coefficients a1, …, an are integers.) Their work introduced the important concept of an ideal in such rings, so called because it could be represented by "ideal elements" outside the ring concerned. In the late 19th century the German mathematician David Hilbert used ideals to solve an old problem about polynomials (algebraic expressions using many variables x1, x2, x3, …). The problem was to take a finite number of variables and decide which ideals could be generated by at most finitely many polynomials. Hilbert's method solved the problem and brought an end to further investigation by showing that they all had this property. His abstract "hands off" approach led the German mathematician Paul Gordon to exclaim “Das ist nicht Mathematik, das ist Theologie!” (“That is not mathematics, that is theology!”). The power of modern algebra had arrived.

Rings can arise naturally in solving mathematical problems, as shown in the following example: Which whole numbers can be written as the sum of two squares? In other words, when can a whole number n be written as a2 + b2? To solve this problem, it is useful to factor n into prime factors, and it is also useful to factor a2 + b2 as (a + bi)(a - bi), where i2 = -1. The question can then be rephrased in terms of numbers a + bi where a and b are integers. This set of numbers forms a ring, and, by considering factorization in this ring, the original problem can be solved. Rings of this sort are very useful in number theory.

Rings in algebraic geometry

Rings are used extensively in algebraic geometry. Consider a curve in the plane given by an equation in two variables such as y2 = x3 + 1. The curve shown in the figure consists of all points (x, y) that satisfy the equation. For example, (2, 3) and (-1, 0) are points on the curve. Every algebraic function in two variables assigns a value to every point of the curve. For example, xy + 2x assigns the value 10 to the point (2, 3) and -2 to the point (-1, 0). Such functions can be added and multiplied together, and they form a ring that can be used to study the original curve. Functions such as y2 and x3 + 1 that agree with each other at every point of the curve are treated as the same function, and this allows the curve to be recovered from the ring. Geometric problems can therefore be transformed into algebraic problems, solved using techniques from modern algebra, and then transformed back into geometric results.

The development of these methods for the study of algebraic geometry was one of the major advances in mathematics during the 20th century. Pioneering work in this direction was done in France by the mathematicians André Weil in the 1950s and Alexandre Grothendieck in the 1960s.

History

Dedekind

The study of rings originated from the theory of polynomial rings and the theory of algebraic integers. In 1871, Richard Dedekind defined the concept of the ring of integers of a number field. In this context, he introduced the terms “ideal” (inspired by Ernst Kummer’s notion of ideal number) and “module” and studied their properties. Dedekind did not use the term “ring” and did not define the concept of a ring in a general setting.

Hilbert

The term "Zahlring" (number ring) was coined by David Hilbert in 1892 and published in 1897. In 19th century German, the word "Ring" could mean "association", which is still used today in English in a limited sense (for example, spy ring), so if that were the etymology then it would be similar to the way "group" entered mathematics by being a non-technical word for "collection of related things". According to Harvey Cohn, Hilbert used the term for a ring that had the property of "circling directly back" to an element of itself (in the sense of an equivalence). Specifically, in a ring of algebraic integers, all high powers of an algebraic integer can be written as an integral combination of a fixed set of lower powers, and thus the powers "cycle back". For instance, if a3 - 4a + 1 = 0 then a3 = 4a - 1, a4 = 4a2 - a, a5 = -a2 + 16a - 4, a6 = 16a2 - 8a + 1, a7 = -8a2 + 65a - 16, and so on; in general, an is going to be an integral linear combination of 1, a, and a2.

Fraenkel and Noether

The first axiomatic definition of a ring was given by Adolf Fraenkel in 1915, but his axioms were stricter than those in the modern definition. For instance, he required every non-zero-divisor to have a multiplicative inverse. In 1921, Emmy Noether gave a modern axiomatic definition of commutative rings (with and without 1) and developed the foundations of commutative ring theory in her paper Idealtheorie in Ringbereichen.

Multiplicative identity and the term "ring"

Fraenkel's axioms for a "ring" included that of a multiplicative identity, whereas Noether's did not.

Most or all books on algebra up to around 1960 followed Noether's convention of not requiring a 1 for a "ring". Starting in the 1960s, it became increasingly common to see books including the existence of 1 in the definition of "ring", especially in advanced books by notable authors such as Artin, Atiyah and MacDonald, Bourbaki, Eisenbud, and Lang. There are also books published as late as 2006 that use the term without the requirement for a 1.

Gardner and Wiegandt assert that, when dealing with several objects in the category of rings (as opposed to working with a fixed ring), if one requires all rings to have a 1, then some consequences include the lack of existence of infinite direct sums of rings, and that proper direct summands of rings are not subrings. They conclude that "in many, maybe most, branches of ring theory the requirement of the existence of a unity element is not sensible, and therefore unacceptable." Poonen makes the counterargument that rings without a multiplicative identity are not totally associative (the product of any finite sequence of ring elements, including the empty sequence, is well-defined, independent of the order of operations) and writes "the natural extension of associativity demands that rings should contain an empty product, so it is natural to require rings to have a 1".

Authors who follow either convention for the use of the term "ring" may use one of the following terms to refer to objects satisfying the other convention:

to include a requirement a multiplicative identity: "unital ring", "unitary ring", "unit ring", "ring with unity", "ring with identity", "ring with a unit", or "ring with 1".

to omit a requirement for a multiplicative identity: "rng" or "pseudo-ring", although the latter may be confusing because it also has other meanings.

COMMUTATIVE RINGS

A ring is called commutative if its multiplication is commutative. Commutative rings resemble familiar number systems, and various definitions for commutative rings are designed to formalize properties of the integers. Commutative rings are also important in algebraic geometry. In commutative ring theory, numbers are often replaced by ideals, and the definition of the prime ideal tries to capture the essence of prime numbers. Integral domains, non-trivial commutative rings where no two non-zero elements multiply to give zero, generalize another property of the integers and serve as the proper realm to study divisibility. Principal ideal domains are integral domains in which every ideal can be generated by a single element, another property shared by the integers. Euclidean domains are integral domains in which the Euclidean algorithm can be carried out. Important examples of commutative rings can be constructed as rings of polynomials and their factor rings. Summary: Euclidean domain =>principal ideal domain =>unique factorization domain =>integral domain =>Commutative ring.

Algebraic geometry

Algebraic geometry is in many ways the mirror image of commutative algebra. This correspondence started with Hilbert's Nullstellensatz that establishes a one-to-one correspondence between the points of an algebraic variety, and the maximal ideals of its coordinate ring. This correspondence has been enlarged and systematized for translating (and proving) most geometrical properties of algebraic varieties into algebraic properties of associated commutative rings.Alexander Grothendieck completed this by introducing schemes, a generalization of algebraic varieties, which may be built from any commutative ring. More precisely, the spectrum of a commutative ring is the space of its prime ideals equipped with

Zariski topology, and augmented with a sheaf of rings. These objects are the "affine schemes" (generalization of affine varieties), and a general scheme is then obtained by "gluing together" (by purely algebraic methods) several such affine schemes, in analogy to the way of constructing a manifold by gluing together the charts of an atlas.

Algebraic geometry is a branch of mathematics, classically studying zeros of multivariate polynomials. Modern algebraic geometry is based on the use of abstract algebraic techniques, mainly from commutative algebra, for solving geometrical problems about these sets of zeros. The fundamental objects of study in algebraic geometry are algebraic varieties, which are geometric manifestations of solutions of systems of polynomial equations. Examples of the most studied classes of algebraic varieties are: plane algebraic curves, which include lines, circles, parabolas, ellipses, hyperbolas, cubic curves like elliptic curves, and quartic curves like lemniscates and Cassini ovals. A point of the plane belongs to an algebraic curve if its coordinates satisfy a given polynomial equation. Basic questions involve the study of the points of special interest like the singular points, the inflection points and the points at infinity. More advanced questions involve the topology of the curve and relations between the curves given by different equations.

Algebraic geometry occupies a central place in modern mathematics and has multiple conceptual connections with such diverse fields as complex analysis, topology and number theory. Initially a study of systems of polynomial equations in several variables, the subject of algebraic geometry starts where equation solving leaves off, and it becomes even more important to understand the intrinsic properties of the totality of solutions of a system of equations, than to find a specific solution; this leads into some of the deepest areas in all of mathematics, both conceptually and in terms of technique. In the 20th century, algebraic geometry split into several subareas.

- The mainstream of algebraic geometry is devoted to the study of the complex points of the algebraic varieties and more generally to the points with coordinates in an algebraically closed field.
- Real algebraic geometry is the study of the real points of an algebraic variety.
- Diophantine geometry and, more generally, arithmetic geometry is the study of the points of an algebraic variety with coordinates in fields that are not algebraically closed and occur in algebraic number theory, such as the field of rational numbers, number fields, finite fields, function fields, and p-adic fields.

- A large part of singularity theory is devoted to the singularities of algebraic varieties.
- Computational algebraic geometry is an area that has emerged at the intersection of algebraic geometry and computer algebra, with the rise of computers. It consists mainly of algorithm design and software development for the study of roperties of explicitly given algebraic varieties.

Much of the development of the mainstream of algebraic geometry in the 20th century occurred within an abstract algebraic framework, with increasing emphasis being placed on "intrinsic" properties of algebraic varieties not dependent on any particular way of embedding the variety in an ambient coordinate space; this parallels developments in topology, differential and complex geometry. One key achievement of this abstract algebraic geometry is Grothendieck's scheme theory which allows one to use sheaf theory to study algebraic varieties in a way which is very similar to its use in the study of differential and analytic manifolds. This is obtained by extending the notion of point.

In classical algebraic geometry, a point of an affine variety may be identified, through Hilbert's Nullstellensatz, with a maximal ideal of the coordinate ring, while the points of the corresponding affine scheme are all prime ideals of this ring. This means that a point of such a scheme may be either a usual point or a subvariety. This approach also enables a unification of the language and the tools of classical algebraic geometry, mainly concerned with complex points, and of algebraic number theory. Wiles' proof of the longstanding conjecture called Fermat's last theorem is an example of the power of this approach.

BASIC NOTIONS

Zeros of simultaneous polynomials

In classical algebraic geometry, the main objects of interest are the vanishing sets of collections of polynomials, meaning the set of all points that simultaneously satisfy one or more polynomial equations. For instance, the two-dimensional sphere of radius 1 in three-dimensional Euclidean space R3 could be defined as the set of all points (x,y,z) with $x^2 + y^2 + z^2 + 1=0$

A "slanted" circle in R3 can be defined as the set of all points (x,y,z) which satisfy the two polynomial equations $x^2 + y^2 + z^2 -1=0$,

$$x+y+z=0,$$

Affine varieties

First we start with a field k. In classical algebraic geometry, this field was always the complex numbers C, but many of the same results are true if we assume only that k is algebraically closed. We consider the affine space of dimension n over k, denoted An(k) (or more simply An, when k is clear from the context). When one fixes a coordinate system, one may identify An(k) with kn. The purpose of not working with kn is to emphasize that one "forgets" the vector space structure that kn carries.

A function f : An '! A1 is said to be polynomial (or regular) if it can be written as a polynomial, that is, if there is a polynomial p in k[x1,...,xn] such that f(M) = p(t1,...,tn) for every point M with coordinates (t1,...,tn) in An. The property of a function to be polynomial (or regular) does not depend on the choice of a coordinate system in A. When a coordinate system is chosen, the regular functions on the affine n-space may be identified with the ring of polynomial functions in n variables over k. Therefore, the set of the regular functions on An is a ring, which is denoted k[An]. We say that a polynomial vanishes at a point if evaluating it at that point gives zero. Let S be a set of polynomials in k[An]. The vanishing set of S (or vanishing locus or zero set) is the set V(S) of all points in An where every polynomial in S vanishes. Symbolically,

$$V(S) = \{(t_1,..,t_n) | \, p(t_1,...,t_n) = 0 \text{ for all } p \in S\}$$

A subset of An which is V(S), for some S, is called an algebraic set. The V stands for variety (a specific type of algebraic set to be defined below). Given a subset U of An, can one recover the set of polynomials which generate it? If U is any subset of An, define I(U) to be the set of all polynomials whose vanishing set contains U. The I stands for ideal: if two polynomials f and g both vanish on U, then f+g vanishes on U, and if h is any polynomial, then hf vanishes on U, so I(U) is always an ideal of the polynomial ring $k[A^n]$.

Two natural questions to ask are:

- Given a subset U of An, when is U = V(I(U))?
- Given a set S of polynomials, when is S = I(V(S))?

The answer to the first question is provided by introducing the Zariski topology, a topology on An whose closed sets are the algebraic sets, and which directly reflects the algebraic structure of k[An]. Then U = V(I(U)) if and only if U is an algebraic set or equivalently a Zariski-closed set. The answer to the second question is given by Hilbert's Nullstellensatz. In one of its forms, it says that I(V(S)) is the radical of the ideal generated by S. In more abstract language, there is a Galois connection, giving rise to two closure operators; they can be identified, and naturally play a basic role in the theory; the example is elaborated at Galois connection.

For various reasons we may not always want to work with the entire ideal corresponding to an algebraic set U. Hilbert's basis theorem implies that ideals in k[An] are always finitely generated. An algebraic set is called irreducible if it cannot be written as the union of two smaller algebraic sets. Any algebraic set is a finite union of irreducible algebraic sets and this decomposition is unique. Thus its elements are called the irreducible components of the algebraic set. An irreducible algebraic set is also called a variety. It turns out that an algebraic set is a variety if and only if it may be defined as the vanishing set of a prime ideal of the polynomial ring. Some authors do not make a clear distinction between algebraic sets and varieties and use irreducible variety to make the distinction when needed.

Regular functions

Just as continuous functions are the natural maps on topological spaces and smooth functions are the natural maps on differentiable manifolds, there is a natural class of functions on an algebraic set, called regular functions or polynomial functions. A regular function on an algebraic set V contained in An is the restriction to V of a regular function on An. For an algebraic set defined on the field of the complex numbers, the regular functions are smooth and even analytic. It may seem unnaturally restrictive to require that a regular function always extend to the ambient space, but it is very similar to the situation in a normal topological space, where the Tietze extension theorem guarantees that a continuous function on a closed subset always extends to the ambient topological space.

Just as with the regular functions on affine space, the regular functions on V form a ring, which we denote by k[V]. This ring is called the coordinate ring of V. Since regular functions on V come from regular functions on An, there is a relationship between the coordinate rings. Specifically, if a regular function on V is the restriction of two functions f and g in k[An], then f " g is a polynomial function which is null on V and thus belongs to I(V). Thus k[V] may be identified with k[An]/I(V).

Morphism of affine varieties

Using regular functions from an affine variety to A1, we can define regular maps from one affine variety to another. First we will define a regular map from a variety into affine space: Let V be a variety contained in An. Choose m regular functions on V, and call them f1,..., fm. We define a regular map f from V to Am by letting f = (f1,..., fm). In other words, each fi determines one coordinate of the range of f. If V2 is a variety contained in Am, we say that f is a regular map from V to V2 if the range of f is contained in V2. The definition of the regular maps apply also to algebraic sets. The regular maps are also called morphisms, as they make the collection of all affine algebraic sets into a category, where the objects are the affine algebraic sets and the morphisms are the regular maps. The affine varieties is a subcategory of the category of the algebraic sets.

Given a regular map g from V to V2 and a regular function f of k[V2], then f "' g " k[V]. The map f '! f "' g is a ring homomorphism from k[V2] to k[V]. Conversely, every ring homomorphism from k[V2] to k[V] defines a regular map from V to V2. This defines an equivalence of categories between the category of algebraic sets and the opposite category of the finitely generated reduced k-algebras. This equivalence is one of the starting points of scheme theory.

Rational function and birational equivalence

In contrast to the preceding sections, this section concerns only varieties and not algebraic sets. On the other hand, the definitions extend naturally to projective varieties (next section), as an affine variety and its projective completion have the same field of functions. If V is an affine variety, its coordinate ring is an integral domain and has thus a field of fractions which is denoted k(V) and called the field of the rational functions on V or, shortly, the function field of V. Its elements are the restrictions to V of the rational functions over the affine space containing V. The domain of a rational function f is not V but the complement of the subvariety (a hypersurface) where the denominator of f vanishes. As with regular maps, one may define a rational map from a variety V to a variety V'. As with the regular maps, the rational maps from V to V' may be identified to the field homomorphisms from k(V') to k(V). Two affine varieties are birationally equivalent if there are two rational functions between them which are inverse one to the other in the regions where both are defined. Equivalently, they are birationally equivalent if their function fields are isomorphic. An affine variety is a rational variety if it is birationally equivalent to an affine space. This means that the variety admits a rational parameterization. For example, the circle of equation $x^2 + y^2$ " $1 = 0$ $x^2 + y^2 - 1=0$ is a rational curve, as it has the parameterization

$$x = \frac{2t}{1+t^2} \quad y = \frac{1-t^2}{1+t^2},$$

which may also be viewed as a rational map from the line to the circle.

The problem of resolution of singularities is to know if every algebraic variety is birationally equivalent to a variety whose projective completion is nonsingular (see also smooth completion). It was solved in the affirmative in characteristic 0 by Heisuke Hironaka in 1964 and is yet unsolved in finite characteristic.

Projective variety

Just as the formulas for the roots of second, third, and fourth degree polynomials suggest extending real numbers to the more algebraically complete setting of the complex numbers, many properties of algebraic varieties suggest extending affine space to a more geometrically complete projective space. Whereas the complex numbers are obtained by adding the number i, a root of the polynomial x2 + 1, projective space is obtained by adding in appropriate points "at infinity", points where parallel lines may meet. To see how this might come about, consider the variety V(y " x2). If we draw it, we get a parabola. As x goes to positive infinity, the slope of the line from the origin to the point (x, x2) also goes to positive infinity. As x goes to negative infinity, the slope of the same line goes to negative infinity.

Compare this to the variety V(y " x3). This is a cubic curve. As x goes to positive infinity, the slope of the line from the origin to the point (x, x3) goes to positive infinity just as before. But unlike before, as x goes to negative infinity, the slope of the same line goes to positive infinity as well; the exact opposite of the parabola. So the behavior "at infinity" of V(y " x3) is different from the behavior "at infinity" of V(y " x2). The consideration of the projective completion of the two curves, which is their prolongation "at infinity" in the projective plane, allows us to quantify this difference: the point at infinity of the parabola is a regular point, whose tangent is the line at infinity, while the point at infinity of the cubic curve is a cusp. Also, both curves are rational, as they are parameterized by x, and the Riemann-Roch theorem implies that the cubic curve must have a singularity, which must be at infinity, as all its points in the affine space are regular.

Thus many of the properties of algebraic varieties, including birational equivalence and all the topological properties, depend on the behavior "at infinity" and so it is natural to study the varieties in projective space. Furthermore, the introduction of projective techniques made many theorems in algebraic geometry simpler and sharper: For example, Bézout's theorem on the number of intersection points between two varieties can be stated in its sharpest form only in projective space. For these reasons, projective space plays a fundamental role in algebraic geometry. Nowadays, the projective space Pn of dimension n is usually defined as the set of the lines passing through a point, considered as the origin, in the affine space of dimension n + 1, or equivalently to the set of the vector lines in a vector space of dimension n + 1. When a coordinate system has been chosen in the space of dimension n + 1, all the points of a line have the same set of coordinates, up to the multiplication by an element of k. This defines the homogeneous coordinates of a point of Pn as a sequence of n + 1 elements of the base field k, defined up to the multiplication by a nonzero element of k (the same for the whole sequence).

A polynomial in n + 1 variables vanishes at all points of a line passing through the origin if and only if it is homogeneous. In this case, one says that the polynomial vanishes at the corresponding point of Pn. This allows us to define a projective algebraic set in Pn as the set V(f1,..., fk), where a finite set of homogeneous polynomials {f1,..., fk} vanishes. Like for affine algebraic sets, there is a bijection between the projective algebraic sets and the reduced homogeneous ideals which define them. The projective varieties are the projective algebraic sets whose defining ideal is prime. In other words, a projective variety is a projective algebraic set, whose homogeneous coordinate ring is an integral domain, the projective coordinates ring being defined as the quotient of the graded ring or the polynomials in n + 1 variables by the homogeneous (reduced) ideal defining the variety. Every projective algebraic set may be uniquely decomposed into a finite union of projective varieties. The only regular functions which may be defined properly on a projective variety are the constant functions. Thus this notion is not used in projective situations. On the other hand, the field of the rational functions or function field is a useful notion, which, similarly to the affine case, is defined as the set of the quotients of two homogeneous elements of the same degree in the homogeneous coordinate ring.

Computational algebraic geometry

One may date the origin of computational algebraic geometry to meeting EUROSAM'79 (International Symposium on Symbolic and Algebraic Manipulation) held at Marseille, France in June 1979. At this meeting,

- Dennis S. Arnon showed that George E. Collins's Cylindrical algebraic decomposition (CAD) allows the computation of the topology of semi-algebraic sets,
- Bruno Buchberger presented the Gröbner bases and his algorithm to compute them,
- Daniel Lazard presented a new algorithm for solving systems of homogeneous polynomial equations with a computational complexity which is essentially polynomial in the expected number of solutions and thus simply exponential in the number of the unknowns. This algorithm is strongly related with Macaulay's multivariate resultant.

Since then, most results in this area are related to one or several of these items either by using or improving one of these algorithms, or by finding algorithms whose complexity is simply exponential in the number of the variables. A body of mathematical theory complementary to symbolic methods called numerical algebraic geometry has been developed over the last several decades. The main computational method is homotopy continuation. This supports, for example, a model of floating point computation for solving problems of algebraic geometry.

1 Gröbner basis

A Gröbner basis is a system of generators of a polynomial ideal whose computation allows the deduction of many properties of the affine algebraic variety defined by the ideal.

Given an ideal I defining an algebraic set V:

- V is empty (over an algebraically closed extension of the basis field), if and only if the Gröbner basis for any monomial ordering is reduced to {1}.
- By means of the Hilbert series one may compute the dimension and the degree of V from any Gröbner basis of I for a monomial ordering refining the total degree.
- If the dimension of V is 0, one may compute the points (finite in number) of V from any Gröbner basis of I (see Systems of polynomial equations).
- A Gröbner basis computation allows one to remove from V all irreducible components which are contained in a given hypersurface.
- A Gröbner basis computation allows one to compute the Zariski closure of the image of V by the projection on the k first coordinates, and the subset of the image where the projection is not proper.
- More generally Gröbner basis computations allow one to compute the Zariski closure of the image and the critical points of a rational function of V into another affine variety.

Gröbner basis computations do not allow one to compute directly the primary decomposition of I nor the prime ideals defining the irreducible components of V, but most algorithms for this involve Gröbner basis computation. The algorithms which are not based on Gröbner bases use regular chains but may need Gröbner bases in some exceptional situations. Gröbner bases are deemed to be difficult to compute. In fact they may contain, in the worst case, polynomials whose degree is doubly exponential in the number of variables and a number of polynomials which is also doubly exponential. However, this is only a worst case complexity, and the complexity bound of Lazard's algorithm of 1979 may frequently apply. Faugère F5 algorithm realizes this complexity, as it may be viewed as an improvement of

Lazard's 1979 algorithm. It follows that the best implementations allow one to compute almost routinely with algebraic sets of degree more than 100. This means that, presently, the difficulty of computing a Gröbner basis is strongly related to the intrinsic difficulty of the problem.

GRÖBNER BASIS

In mathematics, and more specifically in computer algebra, computational algebraic geometry, and computational commutative algebra, a Gröbner basis is a particular kind of generating set of an ideal in a polynomial ring K[x1,..,xn] over a field K. A Gröbner basis allows many important properties of the ideal and the associated algebraic variety to be deduced easily, such as the dimension and the number of zeros when it is finite. Gröbner basis computation is one of the main practical tools for solving systems of polynomial equations and computing the images of algebraic varieties under projections or rational maps. Gröbner basis computation can be seen as a multivariate, non-linear generalization of both Euclid's algorithm for computing polynomial greatest common divisors, and Gaussian elimination for linear systems.

Gröbner bases were introduced in 1965, together with an algorithm to compute them (Buchberger's algorithm), by Bruno Buchberger in his Ph.D. thesis. He named them after his advisor Wolfgang Gröbner. In 2007, Buchberger received the Association for Computing Machinery's Paris Kanellakis Theory and Practice Award for this work. However, the Russian mathematician Nikolai Günther had introduced a similar notion in 1913, published in various Russian mathematical journals. These papers were largely ignored by the mathematical community until their rediscovery in 1987 by Bodo Renschuch et al. An analogous concept for multivariate power series was developed independently by Heisuke Hironaka in 1964, who named them standard bases. This term has been used by some authors to also denote Gröbner bases. The theory of Gröbner bases has been extended by many authors in various directions. It has been generalized to other structures such as polynomials over principal ideal rings or polynomial rings, and also some classes of non-commutative rings and algebras, like Ore algebras.

Polynomial ring

Gröbner bases are primarily defined for ideals in a polynomial ring R = K [x 1, …, x n] over a field K. Although the theory works for any field, most Gröbner basis computations are done either when K is the field of rationals or the integers modulo a prime number. A polynomial is a sum c 1 M 1 + ï" + c m M m where the ci are nonzero elements of K and the M i are monomials or "power products" of the variables. This means that a monomial M is a product M = x 1 a 1 ï” x n a n, where the a i are nonnegative integers. The vector A = [a 1, …, a n] is called the exponent vector of M. The notation is often abbreviated as x 1 a 1 ï” x n a n = X A. Monomials are uniquely defined by their exponent vectors so computers can represent monomials efficiently as exponent vectors, and polynomials as lists of exponent vectors and their coefficients.

If F = {f 1, …, f k}, is a finite set of polynomials in a polynomial ring R, the ideal generated by F is the set of linear combinations of elements from F with coefficients in all of R: è’ f 1, …, f k é’ = {“ i = 1 k g i f i | g 1, …, g k “ K [x 1, …, x n]}.

Monomial ordering

All operations related to Gröbner bases require the choice of a total order on the monomials, with the following properties of compatibility with multiplication. For all monomials M, N, P,

1. M d” N ú’ M P d” N P {M\leq N\Longleftrightarrow MP\leq NP}
2. M d” M P {M\leq MP}.

A total order satisfying these condition is sometimes called an admissible ordering.

These conditions imply Noetherianity, which means that every strictly decreasing sequence of monomials is finite.

Although Gröbner basis theory does not depend on a particular choice of an admissible monomial ordering, three monomial orderings are specially important for the applications:

- Lexicographical ordering, commonly called lex or plex (for pure lexical ordering).
- Total degree reverse lexicographical ordering, commonly called degrevlex.
- Elimination ordering, lexdeg.

Gröbner basis theory was initially introduced for the lexicographical ordering. It was soon realised that the Gröbner basis for degrevlex is almost always much easier to compute, and that it is almost always easier to compute a lex Gröbner basis by first computing the degrevlex basis and then using a “change of ordering algorithm”. When elimination is needed, degrevlex is not convenient; both lex and lexdeg may be used but, again, many computations are relatively easy with lexdeg and almost impossible with lex. Once a monomial ordering is fixed, the terms of a polynomial (product of a monomial with its nonzero coefficient) are naturally ordered by decreasing monomials (for this order). This makes the representation of a polynomial as an ordered list of pairs coefficient–exponent vector a canonical representation of the polynomials. The first (greatest) term of a polynomial p for this ordering and the corresponding monomial and coefficient are respectively called the leading term, leading monomial and leading coefficient and denoted, in this article, lt(p), lm(p) and lc(p).

Reduction

The concept of reduction, also called multivariate division or normal form computation, is central to Gröbner basis theory. It is a multivariate generalization of the Euclidean division of univariate polynomials. In this section we suppose a fixed monomial ordering, which will not be defined explicitly. Given two polynomials f and g, one says that f is reducible by g if some monomial m in f is a multiple of the leading monomial lm(g) of g. If m happens to be the leading monomial of f then one says that f is lead-reducible by g. If c is the coefficient of m in f and m = q lm(g), the one-step reduction of f by g is the operation that associates to f the polynomial red 1 a (f, g) = f " c lc a (g) q g. {operatorname {red} _{1}(f,g)=f-{\frac {c}{\operatorname {lc} (g)}}\,q\,g.} The main properties of this operation are that the resulting polynomial does not contain the monomial m and that the monomials greater than m (for the monomial ordering) remain unchanged. This operation is not, in general, uniquely defined; if several monomials in f are multiples of lm(g), then one may choose arbitrarily which one to reduce. In practice, it is better to choose the greatest one for the monomial ordering, because otherwise subsequent reductions could reintroduce the monomial that has just been removed.

Given a finite set G of polynomials, one says that f is reducible or lead-reducible by G if it is reducible or lead-reducible, respectively, by an element g of G. If it is the case, then one defines red 1 a (f, G) = red 1 a (f, g) {operatorname {red} _{1}(f,G)=\operatorname {red} _{1}(f,g)}. The (complete) reduction of f by G consists in applying iteratively this operator red 1 {operatorname {red} _{1}} until getting a polynomial red a (f, G) {operatorname {red} (f,G)}, which is irreducible by G. It is called a normal form of f by G. In general this form is not uniquely defined (this is not a canonical form); this non-uniqueness is the starting point of Gröbner basis theory.

For Gröbner basis computations, except at the end, it is not necessary to do a complete reduction: a lead-reduction is sufficient, which saves a large amount of computation. The definition of the reduction shows immediately that, if h is a normal form of f by G, then we have f = h + " g " G q g g, {f=h+\sum _{g\in G}q_{g}\,g,} where the q g {q_{g}} are polynomials. In the case of univariate polynomials, if G consists of a single element g, then h is the remainder of the Euclidean division of f by g, qg is the quotient and the division algorithm is exactly the process of lead-reduction. For this reason, some authors use the term multivariate division instead of reduction.

Formal definition

A Gröbner basis G of an ideal I in a polynomial ring R over a field is a generating set of I characterized by any one of the following properties, stated relatively to some monomial order:

- The ideal generated by the leading terms of polynomials in I equals the ideal generated by the leading terms of G;
- The leading term of any polynomial in I is divisible by the leading term of some polynomial in G;
- The multivariate division of any polynomial in the polynomial ring R by G gives a unique remainder;
- The multivariate division by G of any polynomial in the ideal I gives the remainder 0.

All these properties are equivalent; different authors use different definitions depending on the topic they choose. The last two properties allow calculations in the factor ring R/I with the same facility as modular arithmetic. It is a significant fact of commutative algebra that Gröbner bases always exist, and can be effectively computed for any ideal given by a finite generating subset.

Multivariate division requires a monomial ordering, the basis depends on the monomial ordering chosen, and different orderings can give rise to radically different Gröbner bases. Two of the most commonly used orderings are lexicographic ordering, and degree reverse lexicographic order (also called graded reverse lexicographic order or simply total degree order). Lexicographic order eliminates variables, however the resulting Gröbner bases are often very large and expensive to compute. Degree reverse lexicographic order typically provides for the fastest Gröbner basis computations. In this order monomials are compared first by total degree, with ties broken by taking the smallest monomial with respect to lexicographic ordering with the variables reversed.

In most cases (polynomials in finitely many variables with complex coefficients or, more generally, coefficients over any field, for example), Gröbner bases exist for any monomial ordering. Buchberger's algorithm is the oldest and most well-known method for computing them. Other methods are the Faugère's F4 and F5 algorithms, based on the same mathematics as the Buchberger algorithm, and involutive approaches, based on ideas from differential algebra. There are also three algorithms for converting a Gröbner basis with respect to one monomial order to a Gröbner basis with respect to a different monomial order: the FGLM algorithm, the Hilbert Driven Algorithm and the Gröbner walk algorithm. These algorithms are often employed to compute (difficult) lexicographic Gröbner bases from (easier) total degree Gröbner bases.

A Gröbner basis is termed reduced if the leading coefficient of each element of the basis is 1 and no monomial in any element of the basis is in the ideal generated by the leading terms of the other elements of the basis. In the worst case, computation of a Gröbner basis may require time that is exponential or even doubly exponential in the number of solutions of the polynomial system (for degree reverse lexicographic order and lexicographic order, respectively). Despite these complexity bounds, both standard and reduced Gröbner bases are often computable in practice, and most computer algebra systems contain routines to do so.

The concept and algorithms of Gröbner bases have been generalized to submodules of free modules over a polynomial ring. In fact, if L is a free module over a ring R, then one may consider R•”L as a ring by defining the product of two elements of L to be 0. If R is a polynomial ring, this reduces the theory and the algorithms of Gröbner bases of modules to the theory and the algorithms of Gröbner bases of ideals.

The concept and algorithms of Gröbner bases have also been generalized to ideals over various rings, commutative or not, like polynomial rings over a principal ideal ring or Weyl algebras.

Example and counterexample

The zeroes of f(x,y) form the red parabola; the zeroes of g(x,y) form the three blue vertical lines. Their intersection consists of three points. Let R = Q[x,y] be the ring of bivariate polynomials with rational coefficients and consider the ideal I = <f,g> generated by the polynomials

f(x,y) = x2 - y

g(x,y) = x3 - x

Two other elements of I are the polynomials

k(x,y) = -xf(x,y) + g(x,y) = xy - x

h(x,y) = xk(x,y) “ (y - 1)f(x,y) = y2 - y

Under lexicographic ordering with x > y we have

lt(f) = x2

lt(g) = x3

lt(h) = y2

The ideal generated by {lt(f),lt(g)} only contains polynomials that are divisible by x2 which excludes lt(h) = y2; it follows that {f, g} is not a Gröbner basis for I. On the other hand, we can show that {f, k, h} is indeed a Gröbner basis for I. First note that f and g, and therefore also h, k and all the other polynomials in the ideal I have the following three zeroes in the (x,y) plane in common, as indicated in the figure: {(1,1),(-1,1),(0,0)}. Those three points are not collinear, so I does not contain any polynomial of the first degree. Neither can I contain any polynomials of the special form

$$m(x,y) = cx + p(y)$$

with c a nonzero rational number and p a polynomial in the variable y only; the reason being that such an m can never have two distinct zeroes with the same value for y (in this case, the points (1,1) and (-1,1)).

From the above it follows that I, apart from the zero polynomial, only contains polynomials whose leading term has degree greater than or equal to 2; therefore their leading terms are divisible by at least one of the three monomials

{x2, xy, y2} = {lt(f),lt(k),lt(h)}.

This means that {f, k, h} is a Gröbner basis for I with respect to lexicographic ordering with x > y.

PROPERTIES AND APPLICATIONS OF GRÖBNER BASES

Unless explicitly stated, all the results that follow are true for any monomial ordering (see that article for the definitions of the different orders that are mentioned below). It is a common misconception that the lexicographical order is needed for some of these results. On the contrary, the lexicographical order is, almost always, the most difficult to compute, and using it makes impractical many computations that are relatively easy with graded reverse lexicographic order (grevlex), or, when elimination is needed, the elimination order (lexdeg) which restricts to grevlex on each block of variables.

Equality of ideals

Reduced Gröbner bases are unique for any given ideal and any monomial ordering. Thus two ideals are equal if and only if they have the same (reduced) Gröbner basis (usually a Gröbner basis software always produces reduced Gröbner bases).

Membership and inclusion of ideals

The reduction of a polynomial f by the Gröbner basis G of an ideal I yields 0 if and only if f is in I. This allows to test the membership of an element in an ideal. Another method consists in verifying that the Gröbner basis of G*"{f} is equal to G. To test if the ideal I generated by f1,...,fk is contained in the ideal J, it suffices to test that every fi is in J. One may also test the equality of the reduced Gröbner bases of J and J*"{f1,...,fk}.

Solutions of a system of algebraic equations

Any set of polynomials may be viewed as a system of polynomial equations by equating the polynomials to zero. The set of the solutions of such a system depends only on the generated ideal, and, therefore does not change when the given generating set is replaced by the Gröbner basis, for any ordering, of the generated ideal. Such a solution, with coordinates in an algebraically closed field containing the coefficients of the polynomials, is called a zero of the ideal. In the usual case of rational coefficients, this algebraically closed field is chosen as the complex field. An ideal does not have any zero (the system of equations is inconsistent) if and only if 1 belongs to the ideal (this is Hilbert's Nullstellensatz), or, equivalently, if its Gröbner basis (for any monomial ordering) contains 1, or, also, if the corresponding reduced Gröbner basis is.

Given the Gröbner basis G of an ideal I, it has only a finite number of zeros, if and only if, for each variable x, G contains a polynomial with a leading monomial that is a power of x (without any other variable appearing in the leading term). If it is the case the number of zeros, counted with multiplicity, is equal to the number of monomials that are not multiple of any leading monomial of G. This number is called the degree of the ideal. When the number of zeros is finite, the Gröbner basis for a lexicographical monomial ordering provides, theoretically, a solution: the first coordinates of a solution is a root of the greatest common divisor of polynomials of the basis that depends only of the first variable. After substituting this root in the basis, the second coordinates of this solution is a root of the greatest common divisor of the resulting polynomials that depends only on this second variable, and so on. This solving process is only theoretical, because it implies GCD computation and root-finding of polynomials with approximate coefficients, which are not practicable because of numeric instability. Therefore, other methods have been developed to solve polynomial systems through Gröbner bases (see System of polynomial equations for more details).

Dimension, degree and Hilbert series

The dimension of an ideal I in a polynomial ring R is the Krull dimension of the ring R/I and is equal to the dimension of the algebraic set of the zeros of I. It is also equal to number of hyperplanes in general position which are needed to have an intersection with the algebraic set, which is a finite number of points. The degree of the ideal and of its associated algebraic set is the number of points of this finite intersection, counted with multiplicity. In particular, the degree of a hypersurface is equal to the degree of its definition polynomial.

Both degree and dimension depends only on the set of the leading monomials of the Gröbner basis of the ideal for any monomial ordering. The dimension is the maximal size of a subset S of the variables such that there is no leading monomial depending only on the variables in S. Thus, if the ideal has dimension 0, then for each variable x there is a leading monomial in the Gröbner basis that is a power of x. Both dimension and degree may be deduced from the Hilbert series of the ideal, which is the series “ i = 0 “ d i t i {sum _{i=0}^{\infty}d_{i}t^{i}}, where d i {d_{i}} is the number of monomials of degree i that are not multiple of any leading monomial in the Gröbner basis. The Hilbert series may be summed into a rational fraction

where d is the dimension of the ideal and P (t) {P(t)} is a polynomial such that P (1) {P(1)} is the degree of the ideal. Although the dimension and the degree do not depend on the choice of the monomial ordering, the Hilbert series and the polynomial P (t) {P(t)} change when one changes of monomial ordering. Most computer algebra systems that provide functions to compute Gröbner bases provide also functions for computing the Hilbert series, and thus also the dimension and the degree.

Elimination

The computation of Gröbner bases for an elimination monomial ordering allows computational elimination theory. This is based on the following theorem. The computation of Gröbner bases for an elimination monomial ordering allows computational elimination theory. This is based on the following theorem. Consider a polynomial ring K [x 1, …, x n, y 1, …, y m] = K [X, Y], {K[x_{1},\ldots,x_{n},y_{1},\ldots,y_{m}]=K[X,Y],} in which the variables are split into two subsets X and Y. Let us also choose an elimination monomial ordering “eliminating” X, that is a monomial ordering for which two monomials are compared by comparing first the X-parts, and, in case of equality only, considering the Y-parts. This implies that a monomial containing an X-variable is greater than every monomial independent of X. If G is a Gröbner basis of an ideal I for this monomial ordering, then G)” K [Y] {G\cap K[Y]} is a Gröbner basis of I)” K [Y] {I\cap K[Y]} (this ideal is often called the elimination ideal). Moreover, G)” K [Y] {G\cap K[Y]} consists exactly of the polynomials of G whose leading terms belong to K[Y] (this makes very easy the computation of G)” K [Y] {G\cap K[Y]}, as only the leading monomials need to be checked).

This elimination property has many applications, some of them are reported in the next sections. Another application, in algebraic geometry, is that elimination realizes the geometric operation of projection of an affine algebraic set into a subspace of the ambient space: with above notation, the (Zariski closure of) the projection of the algebraic set defined by the ideal I into the Y-subspace is defined by the ideal I)” K [Y]. {I\cap K[Y].} The lexicographical ordering such that x 1 > ï” > x n {x_{1}>\cdots >x_{n}} is an elimination ordering for every partition {x 1, …, x k}, {x k + 1, …, x n}. {{x_{1},\ldots,x_{k}\},\{x_{k+1},\ldots,x_{n}\}.} Thus a Gröbner basis for this ordering carries much more information than usually necessary. This may explain why Gröbner bases for the lexicographical ordering are usually the most difficult to compute.

Intersecting ideals

If I and J are two ideals generated respectively by {f1,..., fm} and {g1,..., gk}, then a single Gröbner basis computation produces a Gröbner basis of their intersection I)” J. For this, one introduces a new indeterminate t, and one uses an elimination ordering such that the first block contains only t and the other block contains all the other variables (this means that a monomial containing t is greater than every monomial that do not contain t. With this monomial ordering, a Gröbner basis of I)” J consists in the polynomials that do not contain t, in the Gröbner basis of the ideal

In other words, I)” J is obtained by eliminating t in K. This may be proven by remarking that the ideal K consists in the polynomials (a “ b) t + b {(a-b)t+b} such that a “ I {a\in I} and b “ J {b\in J}. Such a polynomial is independent of t if and only a=b, which means that b “ I)” J. {b\in I\cap J.}

Implicitization of a rational curve

A rational curve is an algebraic curve that has a parametric equation of the form where f i (t) {f_{i}(t)} and g i (t) {g_{i}(t)} are univariate polynomials for 1 d” i d” n. One may (and will) suppose that f i (t) {f_{i}(t)} and g i (t) {g_{i}(t)} are coprime (they have no non-constant common factors). Implicitization consists in computing the implicit equations of such a curve. In case of n = 2, that is for plane curves, this may be computed with the resultant. The implicit equation is the following resultant: Elimination with Gröbner bases allow to implicitize for any value of n, simply by eliminating t in the ideal è’ g 1 x 1 “ f 1, …, g n x 2 “ f n è’. {langle g_{1}x_{1}-f_{1},\ldots,g_{n}x_{2}-f_{n}\rangle.} If n = 2, the result is the same as with the resultant, if the map t ¦! (x 1, x 2) {t\mapsto (x_{1},x_{2})} is injective for almost every t. In the other case, the resultant is a power of the result of the elimination.

Saturation

When modeling a problem by polynomial equations, it is highly frequent that some quantities are supposed to be non-zero, because, if they are zero, the problem becomes very different. For example, when dealing with triangles, many properties become false if the triangle is degenerated, that is if the length of one side is equal to the sum of the lengths of the other sides. In such situations, there is no hope of deducing relevant information from the polynomial system if the degenerate solutions are not dropped out. More precisely, the system of equations defines an algebraic set which may have several irreducible components, and one has to remove the components on which the degeneracy conditions are everywhere zero. This is done by saturating the equations by the degeneracy conditions, which may be done by using the elimination property of Gröbner bases.

Definition of the saturation

The localization of a ring consists in adjoining to it the formal inverses of some elements. This section concerns only the case of a single element, or equivalently a finite number of elements (adjoining the inverses of several elements is equivalent to adjoin the inverse of their product. The localization of a ring R by an element f is the ring R f = R [t] / (1 “ f t), {R_{f}=R[t]/(1-ft),} where t is a new indeterminate representing the inverse of f. The localization of an ideal I of R is the ideal R f I {R_{f}I} of R f. {R_{f}.} When R is a polynomial ring, computing in R f {R_{f}} is not efficient, because of the need to manage the denominators. Therefore, the operation of localization is usually replaced by the operation of saturation.

The saturation with respect to f of an ideal I in R is the inverse image of R f I {R_{f}I} under the canonical map from R to R f. {R_{f}.} It is the ideal I : f “ = {g “ R | (“ k “ N) f k g “ I} {I:f^{\infty}=\{g\in R|(\exists k\in \mathbb {N})f^{k}g\in I\}} consisting in all elements of R whose products by some power of f belongs to I. If J is the ideal generated by I and 1-ft in R[t], then I : f “ = J)” R. {I:f^{\infty}=J\cap R.} It follows that, if R is a polynomial ring, a Gröbner basis computation eliminating t allows to compute a Gröbner basis of the saturation of an ideal by a polynomial.

The important property of the saturation, which ensures that it removes from the algebraic set defined by the ideal I the irreducible components on which the polynomial f is zero is the following: The primary decomposition of I : f “ {I:f^{\infty}} consists in the components of the primary decomposition of I that do not contain any power of f.

Computation of the saturation

A Gröbner basis of the saturation by f of a polynomial ideal generated by a finite set of polynomials F, may be obtained by eliminating t in F *” {1 “ t f}, {F\cup \{1-tf\},} that is by keeping the polynomials independent of t in the Gröbner basis of F *” {1 “ t f} {F\cup \{1-tf\}} for an elimination ordering eliminating t. Instead of using F, one may also start from a Gröbner basis of F. Which method is most efficient depends on the problem. However, if the saturation does not remove any component, that is if the ideal is equal to its saturated ideal, computing first the Gröbner basis of F is usually faster. On the other hand, if the saturation removes some components, the direct computation may be dramatically faster. If one wants to saturate with respect to several polynomials f 1, …, f k {f_{1},\ldots,f_{k}} or with respect to a single polynomial which is a product f = f 1 … f k, {f=f_{1}\ldots f_{k},} there are three ways to proceed which give the same result but may have very different computation times (it depends on the problem which is the most efficient).

- Saturating by f = f 1 … f k {f=f_{1}\ldots f_{k}} in a single Gröbner basis computation.
- Saturating by f 1, {f_{1},} then saturating the result by f 2, {f_{2},} and so on.
- Adding to F or to its Gröbner basis the polynomials 1 “ t 1 f 1, …, 1 “ t k f k, {1-t_{1}f_{1},\ldots,1-t_{k}f_{k},} and eliminating the t i {t_{i}} in a single Gröbner basis computation.

Effective Nullstellensatz

Hilbert’s Nullstellensatz has two versions. The first one asserts that a set of polynomials has an empty set of common zeros in an algebraic closure of the field of the coefficients if and only if 1 belongs to the generated ideal. This is easily tested with a Gröbner basis computation, because 1 belongs to an ideal if and only if 1 belongs to the Gröbner basis of the ideal, for any monomial ordering. The second version asserts that the set of common zeros (in an algebraic closure of the field of the coefficients) of an ideal is contained in the hypersurface of the zeros of a polynomial f, if and only if a power of f belongs to the ideal. This may be tested by a saturating the ideal by f; in fact, a power of f belongs to the ideal if and only if the saturation by f provides a Gröbner basis containing 1.

Implicitization in higher dimension

By definition, an affine rational variety of dimension k may be described by parametric equations of the form where p 0, …, p n {p_{0},\ldots,p_{n}} are n+1 polynomials in the k variables (parameters of the parameterization) t 1, …, t k. {t_{1},\ldots,t_{k}.} Thus the parameters t 1, …, t k {t_{1},\ldots,t_{k}} and the coordinates x 1, …, x n {x_{1},\ldots,x_{n}} of the points of the variety are zeros of the ideal

One could guess that it suffices to eliminate the parameters to obtain the implicit equations of the variety, as it has been done in the case of curves. Unfortunately this is not always the case. If the p i {p_{i}} have a common zero (sometimes called base point), every irreducible component of the non-empty algebraic set defined by the p i {p_{i}} is an irreducible component of the algebraic set defined by I. It follows that, in this case, the direct elimination of the t i {t_{i}} provides an empty set of polynomials. Therefore, if k>1, two Gröbner basis computations are needed to implicitize:

1. Saturate I {I} by p 0 {p_{0}} to get a Gröbner basis G {G}
2. Eliminate the t i {t_{i}} from G {G} to get a Gröbner basis of the ideal (of the implicit equations) of the variety.

Implementations

- CoCoA free computer algebra system for computing Gröbner bases.
- GAP free computer algebra system that can perform Gröbner bases calculations.
- FGb, Faugère's own implementation of his F4 algorithm, available as a Maple library. To the date, as of 2014, it is, with Magma, the fastest implementation for rational coefficients and coefficients in a finite field of prime order.
- Macaulay 2 free software for doing polynomial computations, particularly Gröbner bases calculations.
- Magma has a very fast implementation of the Faugère's F4 algorithm.
- Maple has implementations of the Buchberger and Faugère F4 algorithms, as well as Gröbner trace.

- Mathematica includes an implementation of the Buchberger algorithm, with performance-improving techniques such as the Gröbner walk, Gröbner trace, and an improvement for toric bases.
- SINGULAR free software for computing commutative and noncommutative Gröbner bases.
- SageMath provides a unified interface to several computer algebra systems (including SINGULAR and Macaulay), and includes a few Gröbner basis algorithms of its own.
- SymPy Python computer algebra system uses Gröbner bases to solve polynomial systems.

CYLINDRICAL ALGEBRAIC DECOMPOSITION (CAD)

CAD is an algorithm which was introduced in 1973 by G. Collins to implement with an acceptable complexity the Tarski–Seidenberg theorem on quantifier elimination over the real numbers. This theorem concerns the formulas of the first-order logic whose atomic formulas are polynomial equalities or inequalities between polynomials with real coefficients. These formulas are thus the formulas which may be constructed from the atomic formulas by the logical operators and (∧) or (∨), not (¬), for all (∀) andexists (∃) Tarski's theorem asserts that, from such a formula, one may compute an equivalent formula without quantifier (∀∃) The complexity of CAD is doubly exponential in the number of variables. This means that CAD allows, in theory, to solve every problem of real algebraic geometry which may be expressed by such a formula, that is almost every problem concerning explicitly given varieties and semi-algebraic sets.

While Gröbner basis computation has doubly exponential complexity only in rare cases, CAD has almost always this high complexity. This implies that, unless if most polynomials appearing in the input are linear, it may not solve problems with more than four variables. Since 1973, most of the research on this subject is devoted either to improve CAD or to find alternative algorithms in special cases of general interest. As an example of the state of art, there are efficient algorithms to find at least a point in every connected component of a semi-algebraic set, and thus to test if a semi-algebraic set is empty. On the other hand, CAD is yet, in practice, the best algorithm to count the number of connected components.

Asymptotic complexity vs. practical efficiency

The basic general algorithms of computational geometry have a double exponential worst case complexity. More precisely, if d is the maximal degree of the input polynomials and n the number of variables, their complexity is at most d^2 c n d^2 for some constant c, and, for some inputs, the complexity is at least d 2 c^2 n d^2 for another constant c2. During the last 20 years of 20th century, various algorithms have been introduced to solve specific subproblems with a better complexity. Most of these algorithms have a complexity d O (n^2) $d^{o(n)}$. Among these algorithms which solve a sub problem of the problems solved by Gröbner bases, one may cite testing if an affine variety is empty and solving nonhomogeneous polynomial systems which have a finite number of solutions. Such algorithms are rarely implemented because, on most entries Faugère's F4 and F5 algorithms have a better practical efficiency and probably a similar or better complexity (probably because the evaluation of the complexity of Gröbner basis algorithms on a particular class of entries is a difficult task which has been done only in a few special cases).

The main algorithms of real algebraic geometry which solve a problem solved by CAD are related to the topology of semi-algebraic sets. One may cite counting the number of connected components, testing if two points are in the same components or computing a Whitney stratification of a real algebraic set. They have a complexity of d O (n 2)$d^{o(n)}$, but the constant involved by O notation is so high that using them to solve any nontrivial problem effectively solved by CAD, is impossible even if one could use all the existing computing power in the world. Therefore, these algorithms have never been implemented and this is an active research area to search for algorithms with have together a good asymptotic complexity and a good practical efficiency.

Abstract modern viewpoint

The modern approaches to algebraic geometry redefine and effectively extend the range of basic objects in various levels of generality to schemes, formal schemes, ind-schemes, algebraic spaces, algebraic stacks and so on. The need for this arises already from the useful ideas within theory of varieties, e.g. the formal functions of Zariski can be accommodated by introducing nilpotent elements in structure rings; considering spaces of loops and arcs, constructing quotients by group actions and developing formal grounds for natural intersection theory and deformation theory lead to some of the further extensions. Most remarkably, in late 1950s, algebraic varieties were subsumed into Alexander Grothendieck's concept of a scheme. Their local objects are affine schemes or prime spectra which are locally ringed spaces which form a category which is antiequivalent to the category of commutative unital rings, extending the duality between the category of affine algebraic varieties over a field k, and the category of finitely generated reduced k-algebras. The gluing is along Zariski topology; one can glue within the category of locally ringed spaces, but also, using the Yoneda embedding, within the more abstract category of presheaves of sets over the category of affine schemes.

The Zariski topology in the set theoretic sense is then replaced by a Grothendieck topology. Grothendieck introduced Grothendieck topologies having in mind more exotic but geometrically finer and more sensitive examples than the crude Zariski topology, namely the étale topology, and the two flat Grothendieck topologies: fppf and fpqc; nowadays some other examples became prominent including Nisnevich topology. Sheaves can be furthermore generalized to stacks in the sense of Grothendieck, usually with some additional representability conditions leading to Artin stacks and, even finer, Deligne-Mumford stacks, both often called algebraic stacks. Sometimes other algebraic sites replace the category of affine schemes. For example, Nikolai Durov has introduced commutative algebraic monads as a generalization of local objects in a generalized algebraic geometry. Versions of a tropical geometry, of an absolute geometry over a field of one element and an algebraic analogue of Arakelov's geometry were realized in this setup.

Another formal generalization is possible to universal algebraic geometry in which every variety of algebras has its own algebraic geometry. The term variety of algebras should not be confused with algebraic variety. The language of schemes, stacks and generalizations has proved to be a valuable way of dealing with geometric concepts and became cornerstones of modern algebraic geometry. Algebraic stacks can be further generalized and for many practical questions like deformation theory and intersection theory, this is often the most natural approach. One can extend the Grothendieck site of affine schemes to a higher categorical site of derived affine schemes, by replacing the commutative rings with an infinity category of differential graded commutative algebras, or of simplicial commutative rings or a similar category with an appropriate variant of a Grothendieck topology. One can also replace presheaves of sets by presheaves of simplicial sets (or of infinity groupoids). Then, in presence of an appropriate homotopic machinery one can develop a notion of derived stack as such a presheaf on the infinity category of derived affine schemes, which is satisfying certain infinite categorical version of a sheaf axiom (and to be algebraic, inductively a sequence of representability conditions). Quillen model categories, Segal categories and quasicategories are some of the most often used tools to formalize this yielding the derived algebraic geometry, introduced by the school of Carlos Simpson, including Andre Hirschowitz, Bertrand Toën, Gabrielle Vezzosi, Michel Vaquié and others; and developed further by Jacob Lurie, Bertrand Toën, and Gabrielle Vezzosi. Another (noncommutative) version of derived algebraic geometry, using A-infinity categories has been developed from early 1990s by Maxim Kontsevich and followers.

History

Before the 16th century

Some of the roots of algebraic geometry date back to the work of the Hellenistic Greeks from the 5th century BC. The Delian problem, for instance, was to construct a length x so that the cube of side x contained the same volume as the rectangular box a2b for given sides a and b. Menaechmus (circa 350 BC) considered the problem geometrically by intersecting the pair of plane conics ay = x2 and xy = ab. The later work, in the 3rd century BC, of Archimedes and Apollonius studied more systematically problems on conic sections, and also involved the use of coordinates. The Arab mathematicians were able to solve by purely algebraic means certain cubic equations, and then to interpret the results geometrically. This was done, for instance, by Ibn al-Haytham in the 10th century AD. Subsequently, Persian mathematician Omar Khayyám (born 1048 A.D.) discovered a method for solving cubic equations by intersecting a parabola with a circle and seems to have been the first to conceive a general theory of cubic equations. A few years after Omar Khayyám, Sharaf al-Din al-Tusi's Treatise on equations has been described as "inaugurating the beginning of algebraic geometry".

Renaissance

Such techniques of applying geometrical constructions to algebraic problems were also adopted by a number of Renaissance mathematicians such as Gerolamo Cardano and Niccolò Fontana "Tartaglia" on their studies of the cubic equation. The geometrical approach to construction problems, rather than the algebraic one, was favored by most 16th and 17th century mathematicians, notably Blaise Pascal who argued against the use of algebraic and analytical methods in geometry. The French mathematicians Franciscus Vieta and later René Descartes and Pierre de Fermat revolutionized the conventional way of thinking about construction problems through the introduction of coordinate geometry. They were interested primarily in the properties of algebraic curves, such as those defined by Diophantine equations (in the case of Fermat), and the algebraic reformulation of the classical Greek works on conics and cubics (in the case of Descartes).

During the same period, Blaise Pascal and Gérard Desargues approached geometry from a different perspective, developing the synthetic notions of projective geometry. Pascal and Desargues also studied curves, but from the purely geometrical point of view: the analog of the Greek ruler and compass construction.

Ultimately, the analytic geometry of Descartes and Fermat won out, for it supplied the 18th century mathematicians with concrete quantitative tools needed to study physical problems using the new calculus of Newton and Leibniz. However, by the end of the 18th century, most of the algebraic character of coordinate geometry was subsumed by the calculus of infinitesimals of Lagrange and Euler.

19th and early 20th century

It took the simultaneous 19th century developments of non-Euclidean geometry and Abelian integrals in order to bring the old algebraic ideas back into the geometrical fold. The first of these new developments was seized up by Edmond Laguerre and Arthur Cayley, who attempted to ascertain the generalized metric properties of projective space. Cayley introduced the idea of homogeneous polynomial forms, and more specifically quadratic forms, on projective space. Subsequently, Felix Klein studied projective geometry (along with other types of geometry) from the viewpoint that the geometry on a space is encoded in a certain class of transformations on the space. By the end of the 19th century, projective geometers were studying more general kinds of transformations on figures in projective space. Rather than the projective linear transformations which were normally regarded as giving the fundamental Kleinian geometry on projective space, they concerned themselves also with the higher degree birational transformations. This weaker notion of congruence would later lead members of the 20th century Italian school of algebraic geometry to classify algebraic surfaces up to birational isomorphism.

The second early 19th century development, that of Abelian integrals, would lead Bernhard Riemann to the development of Riemann surfaces. In the same period began the algebraization of the algebraic geometry through commutative algebra. The prominent results in this direction are Hilbert's basis theorem and Hilbert's Nullstellensatz, which are the basis of the connexion between algebraic geometry and commutative algebra, and Macaulay's multivariate resultant, which is the basis of elimination theory. Probably because of the size of the computation which is implied by multivariate resultants, elimination theory was forgotten during the middle of the 20th century until it was renewed by singularity theory and computational algebraic geometry.

20th century

B. L. van der Waerden, Oscar Zariski and André Weil developed a foundation for algebraic geometry based on contemporary commutative algebra, including valuation theory and the theory of ideals. One of the goals was to give a rigorous framework for proving the results of Italian school of algebraic geometry. In particular, this school used systematically the notion of generic point without any precise definition, which was first given by these authors during the 1930s. In the 1950s and 1960s, Jean-Pierre Serre and Alexander Grothendieck recast the foundations making use of sheaf theory. Later, from about 1960, and largely led by Grothendieck, the idea of schemes was worked out, in conjunction with a very refined apparatus of homological techniques. After a decade of rapid development the field stabilized in the 1970s, and new applications were made, both to number theory and to more classical geometric questions on algebraic varieties, singularities, moduli, and formal moduli. An important class of varieties, not easily understood directly from their defining equations, are the abelian varieties, which are the projective varieties whose points form an abelian group. The prototypical examples are the elliptic curves, which have a rich theory. They were instrumental in the proof of Fermat's last theorem and are also used in elliptic-curve cryptography. In parallel with the abstract trend of the algebraic geometry, which is concerned with general statements about varieties, methods for effective computation with concretely-given varieties have also been developed, which lead to the new area of computational algebraic geometry. One of the founding methods of this area is the theory of Gröbner bases, introduced by Bruno Buchberger in 1965. Another founding method, more specially devoted to real algebraic geometry, is the cylindrical algebraic decomposition, introduced by George E. Collins in 1973.

Analytic geometry

An analytic variety is defined locally as the set of common solutions of several equations involving analytic functions. It is analogous to the included concept of real or complex algebraic variety. Any complex manifold is an analytic variety. Since analytic varieties may have singular points, not all analytic varieties are manifolds. Modern analytic geometry is essentially equivalent to real and complex algebraic geometry, as has been shown by Jean-Pierre Serre in his paper GAGA, the name of which is French for Algebraic geometry and analytic geometry. Nevertheless, the two fields remain distinct, as the methods of proof are quite different and algebraic geometry includes also geometry in finite characteristic.

Applications

Algebraic geometry now finds applications in statistics, control theory, robotics, error-correcting codes, phylogenetics and geometric modelling. There are also connections to string theory, game theory, graph matchings, solitons and integer programming.

Noncommutative rings

Noncommutative rings resemble rings of matrices in many respects. Following the model of algebraic geometry, attempts have been made recently at defining noncommutative geometry based on noncommutative rings. Noncommutative rings and associative algebras (rings that are also vector spaces) are often studied via their categories of modules. A module over a ring is an abelian group that the ring acts on as a ring of endomorphisms, very much akin to the way fields (integral domains in which every non-zero element is invertible) act on vector spaces. Examples of noncommutative rings are given by rings of square matrices or more generally by rings of endomorphisms of abelian groups or modules, and by monoid rings.

Noncommutative ring

In mathematics, more specifically abstract algebra and ring theory, a noncommutative ring is a ring whose multiplication is not commutative; that is, there exists a and b in R with a·b “‘ b·a. Many authors use the term noncommutative ring to refer to rings which are not necessarily commutative, and hence include commutative rings in their definition. Noncommutative algebra is the study of results applying to rings that are not required to be commutative. Many important results in the field of noncommutative algebra area apply to commutative rings as special cases. Although some authors do not assume that rings have a multiplicative identity, in this article we make that assumption unless stated otherwise.

History

Beginning with division rings arising from geometry, the study of noncommutative rings has grown into a major area of modern algebra. The theory and exposition of noncommutative rings was expanded and refined in the 19th and 20th centuries by numerous authors. An incomplete list of such contributors includes E. Artin, Richard Brauer, P. M. Cohn, W. R. Hamilton, I. N. Herstein, N. Jacobson, K. Morita, E. Noether, Ø. Ore and others. Differences between commutative and noncommutative algebra Because noncommutative rings are a much larger class of rings than the commutative rings, their structure and behavior is less well understood. A great deal of work has been done successfully generalizing some results from commutative rings to noncommutative rings. A major difference between rings which are and are not commutative is the necessity to separately consider right ideals and left ideals. It is common for noncommutative ring theorists to enforce a condition on one of these types of ideals while not requiring it to hold for the opposite side. For commutative rings, the left–right distinction does not exist.

IMPORTANT CLASSES

Division rings

A division ring, also called a skew field, is a ring in which division is possible. Specifically, it is a nonzero ring in which every nonzero element a has a multiplicative inverse, i.e., an element x with a·x = x·a = 1. Stated differently, a ring is a division ring if and only if the group of units equals the set of all nonzero elements. Division rings differ from fields only in that their multiplication is not required to be commutative. However, by Wedderburn's little theorem all finite division rings are commutative and therefore finite fields. Historically, division rings were sometimes referred to as fields, while fields were called "commutative fields".

Semisimple rings

A module over a (not necessarily commutative) ring with unity is said to be semisimple (or completely reducible) if it is the direct sum of simple (irreducible) submodules. A ring is said to be (left)-semisimple if it is semisimple as a left module over itself. Surprisingly, a left-semisimple ring is also right-semisimple and vice versa. The left/right distinction is therefore unnecessary.

Semisimple module

In mathematics, especially in the area of abstract algebra known as module theory, a semisimple module or completely reducible module is a type of module that can be understood easily from its parts. A ring that is a semisimple module over itself is known as an Artinian semisimple ring. Some important rings, such as group rings of finite groups over fields of characteristic zero, are semisimple rings. An Artinian ring is initially understood via its largest semisimple quotient. The structure of Artinian semisimple rings is well understood by the Artin–Wedderburn theorem, which exhibits these rings as finite direct products of matrix rings.

Definition

A module over a (not necessarily commutative) ring with unity is said to be semisimple (or completely reducible) if it is the direct sum of simple (irreducible) submodules.

For a module M, the following are equivalent:

1. M is semisimple; i.e., a direct sum of irreducible modules.
2. M is the sum of its irreducible submodules.
3. Every submodule of M is a direct summand: for every submodule N of M, there is a complement P such that M = N •" P.

For 3 Ò! 2 {3\Rightarrow 2}, the starting idea is to find an irreducible submodule by picking any nonzero x " M {x\in M} and letting P {P} be a maximal submodule such that x " P {x\notin P}. It can be shown that the complement of P {P} is irreducible. The most basic example of a semisimple module is a module over a field; i.e., a vector space. On the other hand, the ring Z of integers is not a semisimple module over itself (because, for example, it is not an artinian ring.)

Semisimple is stronger than completely decomposable, which is a direct sum of indecomposable submodules. Let A be an algebra over a field k. Then a left module M over A is said to be absolutely semisimple if, for any field extension F of k, F —" k M {F\otimes _{k}M} is a semisimple module over F —" k A {F\otimes _{k}A}.

Properties

- If M is semisimple and N is a submodule, then N and M/N are also semisimple.
- If each M i {M_{i}} is a semisimple module, then so is * i M i {bigoplus _{i} M_{i}}.
- A module M is finitely generated and semisimple if and only if it is Artinian and its radical is zero.

Endomorphism rings

- • A semisimple module M over a ring R can also be thought of as a ring homomorphism from R into the ring of abelian group endomorphisms of M. The image of this homomorphism is a semiprimitive ring, and every semiprimitive ring is isomorphic to such an image.
- The endomorphism ring of a semisimple module is not only semiprimitive, but also von Neumann regular, (Lam 2001, p. 62).

Semisimple rings

A ring is said to be (left)-semisimple if it is semisimple as a left module over itself. Surprisingly, a left-semisimple ring is also right-semisimple and vice versa. The left/right distinction is therefore unnecessary, and one can speak of semisimple rings without ambiguity. A semisimple ring may be characterized in terms of homological algebra: namely, a ring R is semisimple if and only if any short exact sequence of left (or right) R-modules splits. In particular, any module over a semisimple ring is injective and projective. Since "projective" implies "flat", a semisimple ring is a von Neumann regular ring. Semisimple rings are of particular interest to algebraists. For example, if the base ring R is semisimple, then all

R-modules would automatically be semisimple. Furthermore, every simple (left) R-module is isomorphic to a minimal left ideal of R, that is, R is a left Kasch ring. Semisimple rings are both Artinian and Noetherian. From the above properties, a ring is semisimple if and only if it is Artinian and its Jacobson radical is zero. If an Artinian semisimple ring contains a field as a central subring, it is called a semisimple algebra.

Examples

- Semisimple if and only if it is artinian and reduced.
- If k is a field and G is a finite group of order n, then the group ring k [G] {k[G]} is semisimple if and only if the characteristic of k does not divide n. This is Maschke's theorem, an important result in group representation theory.
- By the Artin–Wedderburn theorem, a unital Artinian ring R is semisimple if and only if it is (isomorphic to) M n 1 (D 1) × M n 2 (D 2) × ï” × M n r (D r) {M_{n_{1}}(D_{1})\times M_{n_{2}}(D_{2})\times \dots \times M_{n_{r}}(D_{r})}, where each D i {D_{i}} is a division ring and each n i {n_{i}} is a positive integer, and M n (D) {M_{n}(D)} denotes the ring of n-by-n matrices with entries in D.
- An example of a semisimple non-unital ring is M “ (K) {M_{\infty}(K)}, the row-finite, column-finite, infinite matrices over a field K.

Simple rings

One should beware that despite the terminology, not all simple rings are semisimple. The problem is that the ring may be "too big", that is, not (left/right) Artinian. In fact, if R is a simple ring with a minimal left/right ideal, then R is semisimple. Classic examples of simple, but not semisimple, rings are the Weyl algebras, such as the Q {mathbb {Q}} -algebra which is a simple noncommutative domain. These and many other nice examples are discussed in more detail in several noncommutative ring theory texts, including chapter 3 of Lam's text, in which they are described as nonartinian simple rings. The module theory for the Weyl algebras is well studied and differs significantly from that of semisimple rings.

JACOBSON SEMISIMPLE

A ring is called Jacobson semisimple (or J-semisimple or semiprimitive) if the intersection of the maximal left ideals is zero, that is, if the Jacobson radical is zero. Every ring that is semisimple as a module over itself has zero Jacobson radical, but not every ring with zero Jacobson radical is semisimple as a module over itself. A

J-semisimple ring is semisimple if and only if it is an artinian ring, so semisimple rings are often called artinian semisimple rings to avoid confusion. For example, the ring of integers, Z, is J-semisimple, but not artinian semisimple.

Semiprimitive rings

A semiprimitive ring or Jacobson semisimple ring or J-semisimple ring is a ring whose Jacobson radical is zero. This is a type of ring more general than a semisimple ring, but where simple modules still provide enough information about the ring. Rings such as the ring of integers are semiprimitive, and an artinian semiprimitive ring is just a semisimple ring. Semiprimitive rings can be understood as subdirect products of primitive rings, which are described by the Jacobson density theorem

Simple rings

A simple ring is a non-zero ring that has no two-sided ideal besides the zero ideal and itself. A simple ring can always be considered as a simple algebra. Rings which are simple as rings but not as modules do exist: the full matrix ring over a field does not have any nontrivial ideals (since any ideal of M(n,R) is of the form M(n,I) with I an ideal of R), but has nontrivial left ideals (namely, the sets of matrices which have some fixed zero columns). According to the Artin–Wedderburn theorem, every simple ring that is left or right Artinian is a matrix ring over a division ring. In particular, the only simple rings that are a finite-dimensional vector space over the real numbers are rings of matrices over either the real numbers, the complex numbers, or the quaternions. Any quotient of a ring by a maximal ideal is a simple ring. In particular, a field is a simple ring. A ring R is simple if and only its opposite ring Ro is simple. An example of a simple ring that is not a matrix ring over a division ring is the Weyl algebra.

IMPORTANT THEOREMS

Wedderburn's little theorem

Wedderburn's little theorem states that every finite domain is a field. In other words, for finite rings, there is no distinction between domains, skew-fields and fields. The Artin–Zorn theorem generalizes the theorem to alternative rings: every finite simple alternative ring is a field.

Artin–Wedderburn theorem

The Artin–Wedderburn theorem is a classification theorem for semisimple rings and semisimple algebras. The theorem states that an (Artinian) semisimple ring R is isomorphic to a product of finitely many ni-by-ni matrix rings over division rings Di, for some integers ni, both of which are uniquely determined up to permutation of the index i. In particular, any simple left or right Artinian ring is isomorphic to an n-by-n matrix ring over a division ring D, where both n and D are uniquely determined. As a direct corollary, the Artin–Wedderburn theorem implies that every simple ring that is finite-dimensional over a division ring (a simple algebra) is a matrix ring. This is Joseph Wedderburn's original result. Emil Artin later generalized it to the case of Artinian rings.

Jacobson density theorem

The Jacobson density theorem is a theorem concerning simple modules over a ring R. The theorem can be applied to show that any primitive ring can be viewed as a "dense" subring of the ring of linear transformations of a vector space. This theorem first appeared in the literature in 1945, in the famous paper "Structure Theory of Simple Rings Without Finiteness Assumptions" by Nathan Jacobson. This can be viewed as a kind of generalization of the Artin-Wedderburn theorem's conclusion about the structure of simple Artinian rings. More formally, the theorem can be stated as follows: The Jacobson Density Theorem. Let U be a simple right R-module, D = End(UR), and X ," U a finite and D-linearly independent set. If A is a D-linear transformation on U then there exists r " R such that A(x) = x • r for all x in X.

Nakayama's lemma

Let J(R) be the Jacobson radical of R. If U is a right module over a ring, R, and I is a right ideal in R, then define U·I to be the set of all (finite) sums of elements of the form u·i, where · is simply the action of R on U. Necessarily, U·I is a submodule of U. If V is a maximal submodule of U, then U/V is simple. So U·J(R) is necessarily a subset of V, by the definition of J(R) and the fact that U/V is simple. Thus, if U contains at least one (proper) maximal submodule, U·J(R) is a proper submodule of U. However, this need not hold for arbitrary modules U over R, for U need not contain any maximal submodules. Naturally, if U is a Noetherian module, this holds. If R is Noetherian, and U is finitely generated, then U is a Noetherian module over R, and the conclusion is satisfied. Somewhat remarkable is that the weaker assumption, namely that U is finitely generated as an R-module (and no finiteness assumption on R), is sufficient to guarantee the conclusion. This is essentially the statement of Nakayama's lemma.

Precisely, one has the following.

Nakayama's lemma: Let U be a finitely generated right module over a ring R. If U is a non-zero module, then U·J(R) is a proper submodule of U. A version of the lemma holds for right modules over non-commutative unitary rings R. The resulting theorem is sometimes known as the Jacobson–Azumaya theorem.

Noncommutative localization

Localization is a systematic method of adding multiplicative inverses to a ring, and is usually applied to commutative rings. Given a ring R and a subset S, one wants to construct some ring R* and ring homomorphism from R to R*, such that the image of S consists of units (invertible elements) in R*. Further one wants R* to be the 'best possible' or 'most general' way to do this – in the usual fashion this should be expressed by a universal property. The localization of R by S is usually denoted by S "1R; however other notations are used in some important special cases. If S is the set of the non zero elements of an integral domain, then the localization is the field of fractions and thus usually denoted Frac(R).

Localizing non-commutative rings is more difficult; the localization does not exist for every set S of prospective units. One condition which ensures that the localization exists is the Ore condition. One case for non-commutative rings where localization has a clear interest is for rings of differential operators. It has the interpretation, for example, of adjoining a formal inverse D"1 for a differentiation operator D. This is done in many contexts in methods for differential equations. There is now a large mathematical theory about it, named microlocalization, connecting with numerous other branches. The micro- tag is to do with connections with Fourier theory, in particular.

Morita equivalence

Morita equivalence is a relationship defined between rings that preserves many ring-theoretic properties. It is named after Japanese mathematician Kiiti Morita who defined equivalence and a similar notion of duality in 1958. Two rings R and S (associative, with 1) are said to be (Morita) equivalent if there is an equivalence of the category of (left) modules over R, R-Mod, and the category of (left) modules over S, S-Mod. It can be shown that the left module categories R-Mod and S-Mod are equivalent if and only if the right module categories Mod-R and Mod-S are equivalent. Further it can be shown that any functor from R-Mod to S-Mod that yields an equivalence is automatically additive.

Brauer group

The Brauer group of a field K is an abelian group whose elements are Morita equivalence classes of central simple algebras of finite rank over K and addition is induced by the tensor product of algebras. It arose out of attempts to classify division algebras over a field and is named after the algebraist Richard Brauer. The group may also be defined in terms of Galois cohomology. More generally, the Brauer group of a scheme is defined in terms of Azumaya algebras.

Ore conditions

The Ore condition is a condition introduced by Øystein Ore, in connection with the question of extending beyond commutative rings the construction of a field of fractions, or more generally localization of a ring. The right Ore condition for a multiplicative subset S of a ring R is that for a " R and s " S, the intersection aS)" sR `" ". A domain that satisfies the right Ore condition is called a right Ore domain. The left case is defined similarly.

Goldie's theorem

In mathematics, Goldie's theorem is a basic structural result in ring theory, proved by Alfred Goldie during the 1950s. What is now termed a right Goldie ring is a ring R that has finite uniform dimension (also called "finite rank") as a right module over itself, and satisfies the ascending chain condition on right annihilators of subsets of R. Goldie's theorem states that the semiprime right Goldie rings are precisely those that have a semisimple Artinian right classical ring of quotients. The structure of this ring of quotients is then completely determined by the Artin–Wedderburn theorem. In particular, Goldie's theorem applies to semiprime right Noetherian rings, since by definition right Noetherian rings have the ascending chain condition on all right ideals. This is sufficient to guarantee that a right-Noetherian ring is right Goldie. The converse does not hold: every right Ore domain is a right Goldie domain, and hence so is every commutative integral domain. A consequence of Goldie's theorem, again due to Goldie, is that every semiprime principal right ideal ring is isomorphic to a finite direct sum of prime principal right ideal rings. Every prime principal right ideal ring is isomorphic to a matrix ring over a right Ore domain.

REPRESENTATION THEORY

Representation theory is a branch of mathematics that draws heavily on non-commutative rings. It studies abstractalgebraic structures by representing their elements as linear transformations of vector spaces, and studies modules over these abstract algebraic structures. In essence, a representation makes an abstract algebraic object more concrete by describing its elements by matrices and the algebraic operations in terms of matrix addition and matrix multiplication, which is non-commutative. The algebraic objects amenable to such a description include groups, associative algebras and Lie algebras. The most prominent of these (and historically the first) is the representation theory of groups, in which elements of a group are represented by invertible matrices in such a way that the group operation is matrix multiplication.

Some relevant theorems

General

- Isomorphism theorems for rings
- Nakayama's lemma

Structure theorems

- The Artin–Wedderburn theorem determines the structure of semisimple rings
- The Jacobson density theorem determines the structure of primitive rings
- Goldie's theorem determines the structure of semiprimeGoldie rings
- The Zariski–Samuel theorem determines the structure of a commutative principal ideal ring
- The Hopkins–Levitzki theorem gives necessary and sufficient conditions for a Noetherian ring to be an Artinian ring
- Morita theory consists of theorems determining when two rings have "equivalent" module categories
- Cartan–Brauer–Hua theorem gives insight on the structure of division rings
- Wedderburn's little theorem states that finite domains are fields

Other

- The Skolem–Noether theorem characterizes the automorphisms of simple rings

Structures and invariants of rings

Dimension of a commutative ring

The Krull dimension of a commutative ring R is the supremum of the lengths n of all the increasing chains of prime ideals $\mathfrak{p}_0 \subsetneq \mathfrak{p}_1 \subsetneq \cdots \subsetneq \mathfrak{p}_n$. For example, the polynomial ring $k[t_1,...,t_n]$ over a field k has dimension n. The fundamental theorem in the dimension theory states the following numbers coincide for a noetherian local ring (R, m):

- The Krull dimension of R.

- The minimum number of the generators of the m-primary ideals.
- The dimension of the graded ring $gr_{\mathfrak{m}}(R) = \oplus_{k\geq 0} \mathfrak{m}^k / \mathfrak{m}^{k+1}$ (equivalently, one plus the degree of its Hilbert polynomial).

A commutative ring R is said to be catenary if any pair of prime ideals p ," p 2 $\mathfrak{p} = \mathfrak{p}_0 \subsetneq \cdots \subsetneq \mathfrak{p}_n = \mathfrak{p}'$ can be extended to a chain of prime ideals p⊂p of same finite length such that there is no prime ideal that is strictly contained in two consecutive terms. Practically all noetherian rings that appear in application are catenary. If is a catenary local integral domain, then, by definition,

$$\dim R = \text{ht}\mathfrak{p} + \dim R/\mathfrak{p}$$

where $\text{ht}\mathfrak{p} = \dim R_{\mathfrak{p}}$ is the height of p. It is a deep theorem of Ratliff that the converse is also true. If R is an integral domain that is a finitely generated k-algebra, then its dimension is the transcendence degree of its field of fractions over k. If S is an integral extension of a commutative ring R, then S and R have the same dimension. Closely related concepts are those of depth and global dimension. In general, if R is a noetherian local ring, then the depth of R is less than or equal to the dimension of R. When the equality holds, R is called a Cohen–Macaulay ring. A regular local ring is an example of a Cohen–Macaulay ring. It is a theorem of Serre that R is a regular local ring if and only if it has finite global dimension and in that case the global dimension is the Krull dimension of R. The significance of this is that a global dimension is a homological notion.

MORITA EQUIVALENCE

Two rings R, S are said to be Morita equivalent if the category of left modules over R is equivalent to the category of left modules over S. In fact, two commutative rings which are Morita equivalent must be isomorphic, so the notion does not add anything new to the category of commutative rings.

However, commutative rings can be Morita equivalent to noncommutative rings, so Morita equivalence is coarser than isomorphism. Morita equivalence is especially important in algebraic topology and functional analysis.

Morita equivalence

In abstract algebra, Morita equivalence is a relationship defined between rings that preserves many ring-theoretic properties. It is named after Japanese mathematician Kiiti Morita who defined equivalence and a similar notion of duality in 1958.

Motivation

Rings are commonly studied in terms of their modules, as modules can be viewed as representations of rings. Every ring R has a natural R-module structure on itself where the module action is defined as the multiplication in the ring, so the approach via modules is more general and gives useful information. Because of this, one often studies a ring by studying the category of modules over that ring. Morita equivalence takes this viewpoint to a natural conclusion by defining rings to be Morita equivalent if their module categories are equivalent. This notion is of interest only when dealing with noncommutative rings, since it can be shown that two commutative rings are Morita equivalent if and only if they are isomorphic.

Definition

Two rings R and S (associative, with 1) are said to be (Morita) equivalent if there is an equivalence of the category of (left) modules over R, R-Mod, and the category of (left) modules over S, S-Mod. It can be shown that the left module categories R-Mod and S-Mod are equivalent if and only if the right module categories Mod-R and Mod-S are equivalent. Further it can be shown that any functor from R-Mod to S-Mod that yields an equivalence is automatically additive.

Examples

The ring of n-by-n matrices with elements in R, denoted Mn(R), is Morita-equivalent to R for any n > 0. Notice that this generalizes the classification of simple artinian rings given by Artin–Wedderburn theory. To see the equivalence, notice that if X is a left R-module then Xn is an Mn(R)-module where the module structure is given by matrix multiplication on the left of column vectors from X. This allows the definition of a functor from the category of left R-modules to the category of left Mn(R)-modules. The inverse functor is defined by realizing that for any Mn(R)-module there is a left R-module X such that the Mn(R)-module is obtained from X as described above.

Criteria for equivalence

Equivalences can be characterized as follows: if F:R-Mod '! R-Mod are additive (covariant) functors, then F and G are an equivalence if and only if there is a balanced (S,R)-bimodule P such that SP and PR are finitely generated projective generators and there are natural isomorphisms of the functors F a (") E" P —" R " $F(-) \cong P \otimes_R -$, and of the functors G a (") E" Hom a (S P, ") $G(-) \cong \mathrm{Hom}(_S P, -)$. . Finitely generated projective generators are also sometimes called progenerators for their module category. For every right-exact functor F from the category of left-R modules to the category of left-S modules that commutes with direct sums, a theorem of homological algebra shows that there is a (S,R)-bimodule E such that the functor $E\otimes_{s}-$. Since equivalences are by necessity exact and commute with direct sums, this implies that R and S are Morita equivalent if and only if there are bimodules RMS and SNR such that $E\otimes_{s}-M\otimes_{s}N\cong R$ as (R,R) bimodules and $N\otimes_{r}M\cong S$ as (S,S) bimodules. Moreover, N and M are related via an (S,R) bimodule isomorphism: $N\cong \mathrm{Hom}(M_{s},S_{s})$.

More concretely, two rings R and S are Morita equivalent if and only if $S\cong \mathrm{End}(P_{r})$ for a progenerator module PR, which is the case if and only if $S\cong e\mathbb{M}_{n}(R)e$

(isomorphism of rings) for some positive integer n and full idempotent e in the matrix ring Mn(R). It is known that if R is Morita equivalent to S, then the ring C(R) is isomorphic to the ring C(S), where C(-) denotes the center of the ring, and furthermore R/J(R) is Morita equivalent to S/J(S), where J(-) denotes the Jacobson radical. While isomorphic rings are Morita equivalent, Morita equivalent rings can be nonisomorphic. An easy example is that a division ring D is Morita equivalent to all of its matrix rings Mn(D), but cannot be isomorphic when n > 1. In the special case of commutative rings, Morita equivalent rings are actually isomorphic. This follows immediately from the comment above, for if R is Morita equivalent to S, R = C a (R) E" C a (S) = S $R=\mathrm{C}(R)\cong \mathrm{C}(S)=S$. In fact, if R and S are isomorphic commutative rings, every equivalence between R-Mod and S-Mod arises up to natural isomorphism from an isomorphism between R and S.

Properties preserved by equivalence

Many properties are preserved by the equivalence functor for the objects in the module category. Generally speaking, any property of modules defined purely in terms of modules and their homomorphisms (and not to their underlying elements or ring) is a categorical property which will be preserved by the equivalence functor. For example, if F(-) is the equivalence functor from R-Mod to S-Mod, then the R module M has any of the following properties if and only if the S module F(M) does: injective, projective, flat, faithful, simple, semisimple, finitely generated, finitely presented, Artinian, and Noetherian. Examples of properties not necessarily preserved include being free, and being cyclic. Many ring theoretic properties are stated in terms of their modules, and so these properties are preserved between Morita equivalent rings. Properties shared between equivalent rings are called Morita invariant properties. For example, a ring R is semisimple if and only if all of its modules are semisimple, and since semisimple modules are preserved under Morita equivalence, an equivalent ring S must also have all of its modules semisimple, and therefore be a semisimple ring itself. Sometimes it is not immediately obvious why a property should be preserved. For example, using one standard definition of von Neumann regular ring (for all a in R, there exists x in R such that a = axa) it is not clear that an equivalent ring should also be von Neumann regular. However another formulation is: a ring is von Neumann regular if and only if all of its modules are flat. Since flatness is preserved across Morita equivalence, it is now clear that von Neumann regularity is Morita invariant.

The following properties are Morita invariant:

- Simple, semisimple
- Von Neumann regular
- Right (or left) Noetherian, right (or left) Artinian
- Right (or left) self-injective
- Quasi-Frobenius
- Prime, right (or left) primitive, semiprime, semiprimitive
- Right (or left) (semi-)hereditary
- Right (or left) nonsingular
- Right (or left) coherent
- Semiprimary, right (or left) perfect, semiperfect
- Semilocal

Examples of properties which are not Morita invariant include commutative, local, reduced, domain, right (or left) Goldie, Frobenius, invariant basis number, and Dedekind finite. There are at least two other tests for determining whether or not a ring property is Morita invariant. An element e in a ring R is a full idempotent when e2 = e and ReR = R.

- $\mathcal{P}$ is Morita invariant if and only if whenever a ring R satisfies $\mathcal{P}$, then so does eRe for every full idempotent e and so does every matrix ring Mn(R) for every positive integer n;

or

- $\mathcal{P}$ is Morita invariant if and only if: for any ring R and full idempotent e in R, R satisfies $\mathcal{P}$ if and only if the ring eRe satisfies $\mathcal{P}$.

Further directions

Dual to the theory of equivalences is the theory of dualities between the module categories, where the functors used are contravariant rather than covariant. This theory, though similar in form, has significant differences because there is no duality between the categories of modules for any rings, although dualities may exist for subcategories. In other words, because infinite dimensional modules are not generally reflexive, the theory of dualities applies more easily to finitely generated algebras over noetherian rings. Perhaps not surprisingly, the criterion above has an analogue for dualities, where the natural isomorphism is given in terms of the hom functor rather than the tensor functor. Morita equivalence can also be defined in more structured situations, such as for symplectic groupoids and C*-algebras. In the case of C*-algebras, a stronger type equivalence, called strong Morita equivalence, is needed to obtain results useful in applications, because of the additional structure of C*-algebras (coming from the involutive *-operation) and also because C*-algebras do not necessarily have an identity element.

Significance in K-theory

If two rings are Morita equivalent, there is an induced equivalence of the respective categories of projective modules since the Morita equivalences will preserve exact sequences (and hence projective modules). Since the algebraic K-theory of a ring is defined (in Quillen's approach) in terms of the homotopy groups of (roughly) the classifying space of the nerve of the (small) category of finitely generated projective modules over the ring, Morita equivalent rings must have isomorphic K-groups. Finitely generated projective module over a ring and Picard group Let R be a commutative ring and P(R) the set of isomorphism classes of finitely generated projective modules over R; let also $P_n(R)$ subsets consisting of those with constant rank n. (The rank of a module M is the continuous function $Spec R \to \mathbb{Z}, \mathfrak{p} \mapsto \dim M \otimes_R k(\mathfrak{p})$. $P_1(R)$ is usually denoted by Pic(R). It is an abelian group called the Picard group of R. If R is an integral domain with the field of fractions F of R, then there is an exact sequence of groups:

$$1 \to R^* \to F^* \overset{f \mapsto fR}{\to} Cart(R) \to Pic(R) \to 1$$

whereCart a (R) is the set of fractional ideals of R. If R is a regular domain (i.e., regular at any prime ideal), then Pic(R) is precisely the divisor class group of R. For example, if R is a principal ideal domain, then Pic(R) vanishes. In algebraic number theory, R will be taken to be the ring of integers, which is Dedekind and thus regular. It follows that Pic(R) is a finite group (finiteness of class number) that measures the deviation of the ring of integers from being a PID. One can also consider the group completionof P(R); this results in a commutative ring K0(R). Note that K0(R) = K0(S) if two commutative rings R, S are Morita equivalent.

Noncommutative rings

Noncommutative rings resemble rings of matrices in many respects. Following the model of algebraic geometry, attempts have been made recently at defining noncommutative geometry based on noncommutative rings. Noncommutative rings and associative algebras (rings that are also vector spaces) are often studied via their categories of modules. A module over a ring is an abelian group that the ring acts on as a ring of endomorphisms, very much akin to the way fields (integral domains in which every non-zero element is invertible) act on vector spaces. Examples of noncommutative rings are given by rings of square matrices or more generally by rings of endomorphisms of abelian groups or modules, and by monoid rings.

Representation theory

Representation theory is a branch of mathematics that draws heavily on non-commutative rings. It studies abstract algebraic structures by representing their elements as linear transformations of vector spaces, and studies modules over these abstract algebraic structures. In essence, a representation makes an abstract algebraic object more concrete by describing its elements by matrices and the algebraic operations in terms of matrix addition and matrix multiplication, which is non-commutative. The algebraic objects amenable to such a description include groups, associative algebras and Lie algebras. The most prominent of these (and historically the first) is the representation theory of groups, in which elements of a group are represented by invertible matrices in such a way that the group operation is matrix multiplication.

Structure of noncommutative rings

The structure of a noncommutative ring is more complicated than that of a commutative ring. For example, there exist simple rings, containing no non-trivial proper (two-sided) ideals, which contain non-trivial proper left or right ideals. Various invariants exist for commutative rings, whereas invariants of noncommutative rings are difficult to find. As an example, the nilradical of a ring, the set of all nilpotent elements, need not be an ideal unless the ring is commutative. Specifically, the set of all nilpotent elements in the ring of all n x n matrices over a division ring never forms an ideal, irrespective of the division ring chosen. There are, however, analogues of the nilradical defined for noncommutative rings, that coincide with the nilradical when commutativity is assumed.

The concept of the Jacobson radical of a ring; that is, the intersection of all right/left annihilators of simple right/left modules over a ring, is one example. The fact that the Jacobson radical can be viewed as the intersection of all maximal right/left ideals in the ring, shows how the internal structure of the ring is reflected by its modules. It is also a fact that the intersection of all maximal right ideals in a ring is the same as the intersection of all maximal left ideals in the ring, in the context of all rings; whether commutative or noncommutative. Noncommutative rings serve as an active area of research due to their ubiquity in mathematics. For instance, the ring of n-by-nmatrices over a field is noncommutative despite its natural occurrence in geometry, physics and many parts of mathematics. More generally, endomorphism rings of abelian groups are rarely commutative, the simplest example being the endomorphism ring of the Klein four-group. One of the best known noncommutative rings is the division ring of quaternions.

THE RING OF INTEGERS OF A NUMBER FIELD

The coordinate ring of an algebraic variety

If X is an affine algebraic variety, then the set of all regular functions on X forms a ring called the coordinate ring of X. For a projective variety, there is an analogous ring called the homogeneous coordinate ring. Those rings are essentially the same things as varieties: they correspond in essentially a unique way. This may be seen via either Hilbert's Nullstellensatz or scheme-theoretic constructions (i.e., Spec and Proj).

Ring of invariants

A basic (and perhaps the most fundamental) question in the classical invariant theory is to find and study polynomials in the polynomial ring k[V] that are invariant under the action of a finite group (or more generally reductive) G on V. The main example is the ring of symmetric polynomials: symmetric polynomials are polynomials that are invariant under permutation of variable. The fundamental theorem of symmetric polynomials states that this ring is $R[\sigma_1,...,\sigma_n]$ where σ_i are elementary symmetric polynomials.

History

Commutative ring theory originated in algebraic number theory, algebraic geometry, and invariant theory. Central to the development of these subjects were the rings of integers in algebraic number fields and algebraic function fields, and the rings of polynomials in two or more variables. Noncommutative ring theory began with attempts to extend the complex numbers to various hypercomplex number systems. The genesis of the theories of commutative and noncommutative rings dates back to the early 19th century, while their maturity was achieved only in the third decade of the 20th century. More precisely, William Rowan Hamilton put forth the quaternions and biquaternions; James Cockle presented tessarines and coquaternions; and William Kingdon Clifford was an enthusiast of split-biquaternions, which he called algebraic motors. These noncommutative algebras, and the non-associative Lie algebras, were studied within universal algebra before the subject was divided into particular mathematical structure types. One sign of re-organization was the use of direct sums to describe algebraic structure.

The various hypercomplex numbers were identified with matrix rings by Joseph Wedderburn (1908) and Emil Artin (1928). Wedderburn's structure theorems were formulated for finite-dimensional algebras over a field while Artin generalized them to Artinian rings. In 1920, Emmy Noether, in collaboration with W. Schmeidler, published a paper about the theory of ideals in which they defined left and right ideals in a ring. The following year she published a landmark paper called Idealtheorie in Ringbereichen, analyzing ascending chain conditions with regard to (mathematical) ideals. Noted algebraist Irving Kaplansky called this work "revolutionary"; the publication gave rise to the term "Noetherian ring", and several other mathematical objects being called Noetherian.

Differences between commutative and noncommutative algebra

Because noncommutative rings are a much larger class of rings than the commutative rings, their structure and behavior is less well understood. A great deal of work has been done successfully generalizing some results from commutative rings to noncommutative rings. A major difference between rings which are and are not commutative is the necessity to separately consider right ideals and left ideals. It is common for noncommutative ring theorists to enforce a condition on one of these types of ideals while not requiring it to hold for the opposite side. For commutative rings, the left–right distinction does not exist.

Important classes

Division rings

A division ring, also called a skew field, is a ring in which division is possible. Specifically, it is a nonzero ring in which every nonzero element a has a multiplicative inverse, i.e., an element x with $a \cdot x = x \cdot a = 1$. Stated differently, a ring is a division ring if and only if its group of units is the set of all nonzero elements.

Division rings differ from fields only in that their multiplication is not required to be commutative. However, by Wedderburn's little theorem all finite division rings are commutative and therefore finite fields. Historically, division rings were sometimes referred to as fields, while fields were called "commutative fields".

Semisimple rings

A module over a (not necessarily commutative) ring with unity is said to be semisimple (or completely reducible) if it is the direct sum of simple (irreducible) submodules.

A ring is said to be (left)-semisimple if it is semisimple as a left module over itself. Surprisingly, a left-semisimple ring is also right-semisimple and vice versa. The left/right distinction is therefore unnecessary.

Semiprimitive rings

A semiprimitive ring or Jacobson semisimple ring or J-semisimple ring is a ring whose Jacobson radical is zero. This is a type of ring more general than a semisimple ring, but where simple modules still provide enough information about the ring. Rings such as the ring of integers are semiprimitive, and an artinian semiprimitive ring is just a semisimple ring. Semiprimitive rings can be understood as subdirect products of primitive rings, which are described by the Jacobson density theorem.

Simple rings

A simple ring is a non-zero ring that has no two-sided ideal besides the zero ideal and itself. A simple ring can always be considered as a simple algebra. Rings which are simple as rings but not as modules do exist: the full matrix ring over a field does not have any nontrivial ideals (since any ideal of M(n,R) is of the form M(n,I) with I an ideal of R), but has nontrivial left ideals (namely, the sets of matrices which have some fixed zero columns).

According to the Artin–Wedderburn theorem, every simple ring that is left or right Artinian is a matrix ring over a division ring. In particular, the only simple rings that are a finite-dimensional vector space over the real numbers are rings of matrices over either the real numbers, the complex numbers, or the quaternions.

Any quotient of a ring by a maximal ideal is a simple ring. In particular, a field is a simple ring. A ring R is simple if and only its opposite ring Ro is simple.

An example of a simple ring that is not a matrix ring over a division ring is the Weyl algebra.

Important theorems

Wedderburn's little theorem

Wedderburn's little theorem states that every finite domain is a field. In other words, for finite rings, there is no distinction between domains, division rings and fields.

The Artin–Zorn theorem generalizes the theorem to alternative rings: every finite simple alternative ring is a field.

Artin–Wedderburn theorem

The Artin–Wedderburn theorem is a classification theorem for semisimple rings and semisimple algebras. The theorem states that an (Artinian) semisimple ring R is isomorphic to a product of finitely many ni-by-ni matrix rings over division rings Di, for some integers ni, both of which are uniquely determined up to permutation of the index i. In particular, any simple left or right Artinian ring is isomorphic to an n-by-n matrix ring over a division ring D, where both n and D are uniquely determined.

As a direct corollary, the Artin–Wedderburn theorem implies that every simple ring that is finite-dimensional over a division ring (a simple algebra) is a matrix ring. This is Joseph Wedderburn's original result. Emil Artin later generalized it to the case of Artinian rings.

Jacobson density theorem

The Jacobson density theorem is a theorem concerning simple modules over a ring R.

The theorem can be applied to show that any primitive ring can be viewed as a "dense" subring of the ring of linear transformations of a vector space. This theorem first appeared in the literature in 1945, in the famous paper "Structure Theory of Simple Rings Without Finiteness Assumptions" by Nathan Jacobson. This can be viewed as a kind of generalization of the Artin-Wedderburn theorem's conclusion about the structure of simple Artinian rings.

More formally, the theorem can be stated as follows:

The Jacobson Density Theorem. Let U be a simple right R-module, D = End(UR), and X ? U a finite and D-linearly independent set. If A is a D-linear transformation on U then there exists r ? R such that A(x) = x · r for all x in X.

Nakayama's lemma

Let J(R) be the Jacobson radical of R. If U is a right module over a ring, R, and I is a right ideal in R, then define U·I to be the set of all (finite) sums of elements of the form u·i, where · is simply the action of R on U. Necessarily, U·I is a submodule of U.

If V is a maximal submodule of U, then U/V is simple. So U·J(R) is necessarily a subset of V, by the definition of J(R) and the fact that U/V is simple. Thus, if U contains at least one (proper) maximal submodule, U·J(R) is a proper submodule of U. However, this need not hold for arbitrary modules U over R, for U need not contain any maximal submodules. Naturally, if U is a Noetherian module, this holds. If R is Noetherian, and U is finitely generated, then U is a Noetherian module over R, and the conclusion is satisfied. Somewhat remarkable is that the weaker assumption, namely that U is finitely generated as an R-module (and no finiteness assumption on R), is sufficient to guarantee the conclusion. This is essentially the statement of Nakayama's lemma.

Precisely, one has the following.

Nakayama's lemma: Let U be a finitely generated right module over a ring R. If U is a non-zero module, then U·J(R) is a proper submodule of U.

A version of the lemma holds for right modules over non-commutative unitary rings R. The resulting theorem is sometimes known as the Jacobson–Azumaya theorem.

Noncommutative localization

Localization is a systematic method of adding multiplicative inverses to a ring, and is usually applied to commutative rings. Given a ring R and a subset S, one wants to construct some ring R* and ring homomorphism from R to R*, such that the image of S consists of units (invertible elements) in R*. Further one wants R* to be the 'best possible' or 'most general' way to do this – in the usual fashion this should be expressed by a universal property. The localization of R by S is usually denoted by S - 1R; however other notations are used in some important special cases. If S is the set of the non zero elements of an integral domain, then the localization is the field of fractions and thus usually denoted Frac(R).

Localizing non-commutative rings is more difficult; the localization does not exist for every set S of prospective units. One condition which ensures that the localization exists is the Ore condition.

One case for non-commutative rings where localization has a clear interest is for rings of differential operators. It has the interpretation, for example, of adjoining a formal inverse D-1 for a differentiation operator D. This is done in many contexts in methods for differential equations. There is now a large mathematical theory about it, named microlocalization, connecting with numerous other branches. The micro- tag is to do with connections with Fourier theory, in particular.

Morita equivalence

Morita equivalence is a relationship defined between rings that preserves many ring-theoretic properties. It is named after Japanese mathematician Kiiti Morita who defined equivalence and a similar notion of duality in 1958.

Two rings R and S (associative, with 1) are said to be (Morita) equivalent if there is an equivalence of the category of (left) modules over R, R-Mod, and the category of (left) modules over S, S-Mod. It can be shown that the left module categories R-Mod and S-Mod are equivalent if and only if the right module categories Mod-R and Mod-S are equivalent. Further it can be shown that any functor from R-Mod to S-Mod that yields an equivalence is automatically additive.

Brauer group

The Brauer group of a field K is an abelian group whose elements are Morita equivalence classes of central simple algebras of finite rank over K and addition is induced by the tensor product of algebras. It arose out of attempts to classify division algebras over a field and is named after the algebraist Richard Brauer. The group may also be defined in terms of Galois cohomology. More generally, the Brauer group of a scheme is defined in terms of Azumaya algebras.

Ore conditions

The Ore condition is a condition introduced by Øystein Ore, in connection with the question of extending beyond commutative rings the construction of a field of fractions, or more generally localization of a ring. The right Ore condition for a multiplicative subset S of a ring R is that for a ? R and s ? S, the intersection aS n sR ? Ø. A domain that satisfies the right Ore condition is called a right Ore domain. The left case is defined similarly.

Goldie's theorem

In mathematics, Goldie's theorem is a basic structural result in ring theory, proved by Alfred Goldie during the 1950s. What is now termed a right Goldie ring is a ring R that has finite uniform dimension (also called "finite rank") as a right module over itself, and satisfies the ascending chain condition on right annihilators of subsets of R.

Goldie's theorem states that the semiprime right Goldie rings are precisely those that have a semisimple Artinian right classical ring of quotients. The structure of this ring of quotients is then completely determined by the Artin–Wedderburn theorem.

In particular, Goldie's theorem applies to semiprime right Noetherian rings, since by definition right Noetherian rings have the ascending chain condition on all right ideals. This is sufficient to guarantee that a right-Noetherian ring is right Goldie. The converse does not hold: every right Ore domain is a right Goldie domain, and hence so is every commutative integral domain.

A consequence of Goldie's theorem, again due to Goldie, is that every semiprime principal right ideal ring is isomorphic to a finite direct sum of prime principal right ideal rings. Every prime principal right ideal ring is isomorphic to a matrix ring over a right Ore domain.

SOME EXAMPLES OF THE UBIQUITY OF RINGS

Many different kinds of mathematical objects can be fruitfully analyzed in terms of some associated ring.

Cohomology ring of a topological space

To any topological space X one can associate its integral cohomology ring $H^*(X,\mathbf{Z}) = \bigoplus_{i=0}^{\infty} H^i(X,\mathbf{Z})$, a graded ring. There are also homology groups H i (X, Z) of a space, and indeed these were defined first, as a useful tool for distinguishing between certain pairs of topological spaces, like the spheres and tori, for which the methods of point-set topology are not well-suited. Cohomology groups were later defined in terms of homology groups in a way which is roughly analogous to the dual of a vector space. To know each individual integral homology group is essentially the same as knowing each individual integral cohomology group, because of the universal coefficient theorem. However, the advantage of the cohomology groups is that there is a natural product, which is analogous to the observation that one can multiply pointwise a k-multilinear form and an l-multilinear form to get a (k + l)-multilinear form.

The ring structure in cohomology provides the foundation for characteristic classes of fiber bundles, intersection theory on manifolds and algebraic varieties, Schubert calculus and much more.

Burnside ring of a group

To any group is associated its Burnside ring which uses a ring to describe the various ways the group can act on a finite set. The Burnside ring's additive group is the free abelian group whose basis are the transitive actions of the group and whose addition is the disjoint union of the action. Expressing an action in terms of the basis is decomposing an action into its transitive constituents. The multiplication is easily expressed in terms of the representation ring: the multiplication in the Burnside ring is formed by writing the tensor product of two permutation modules as a permutation module. The ring structure allows a formal way of subtracting one action from another. Since the Burnside ring is contained as a finite index subring of the representation ring, one can pass easily from one to the other by extending the coefficients from integers to the rational numbers.

Representation ring of a group ring

To any group ring or Hopf algebra is associated its representation ring or "Green ring". The representation ring's additive group is the free abelian group whose basis are the indecomposable modules and whose addition corresponds to the direct sum. Expressing a module in terms of the basis is finding an indecomposable decomposition of the module. The multiplication is the tensor product. When the algebra is semisimple, the representation ring is just the character ring from character theory, which is more or less the Grothendieck group given a ring structure.

Function field of an irreducible algebraic variety

To any irreducible algebraic variety is associated its function field. The points of an algebraic variety correspond to valuation rings contained in the function field and containing the coordinate ring. The study of algebraic geometry makes heavy use of commutative algebra to study geometric concepts in terms of ring-theoretic properties. Birational geometry studies maps between the subrings of the function field.

Face ring of a simplicial complex

Every simplicial complex has an associated face ring, also called its Stanley–Reisner ring. This ring reflects many of the combinatorial properties of the simplicial complex, so it is of particular interest in algebraic combinatorics. In particular, the algebraic geometry of the Stanley–Reisner ring was used to characterize the numbers of faces in each dimension of simplicial polytopes.

Category-theoretic description

Every ring can be thought of as a monoid in Ab, the category of abelian groups (thought of as a monoidal category under the tensor product of Z-modules). The monoid action of a ring R on an abelian group is simply an R-module. Essentially, an R-module is a generalization of the notion of a vector space – where rather than a vector space over a field, one has a "vector space over a ring".

Let (A, +) be an abelian group and let End(A) be its endomorphism ring (see above). Note that, essentially, End(A) is the set of all morphisms of A, where if f is in End(A), and g is in End(A), the following rules may be used to compute f + g and f · g:

$$(f + g)(x) = f(x) + g(x)$$

$$(f \cdot g)(x) = f(g(x)),$$

where + as in f(x) + g(x) is addition in A, and function composition is denoted from right to left. Therefore, associated to any abelian group, is a ring. Conversely, given any ring, (R, +, ·), (R, +) is an abelian group. Furthermore, for every r in R, right (or left) multiplication by r gives rise to a morphism of (R, +), by right (or left) distributivity. Let A = (R, +). Consider those endomorphisms of A, that "factor through" right (or left) multiplication of R. In other words, let EndR(A) be the set of all morphisms m of A, having the property that m(r · x) = r · m(x). It was seen that every r in R gives rise to a morphism of A: right multiplication by r. It is in fact true that this association of any element of R, to a morphism of A, as a function from R to EndR(A), is an isomorphism of rings. In this sense, therefore, any ring can be viewed as the endomorphism ring of some abelian X-group (by X-group, it is meant a group with X being its set of operators). In essence, the most general form of a ring, is the endomorphism group of some abelian X-group.

Any ring can be seen as a preadditive category with a single object. It is therefore natural to consider arbitrary preadditive categories to be generalizations of rings. And indeed, many definitions and theorems originally given for rings can be translated to this more general context. Additive functors between preadditive categories generalize the concept of ring homomorphism, and ideals in additive categories can be defined as sets of morphisms closed under addition and under composition with arbitrary morphisms.

3

MODULE THEORY IN MODERN ALGEBRA

A module is a mathematical object in which things can be added together commutatively by multiplying coefficients and in which most of the rules of manipulating vectors hold.

A module is abstractly very similar to a vector space, although in modules, coefficients are taken in rings that are much more general algebraic objects than the fields used in vector spaces. A module taking its coefficients in a ring R is called a module over R, or a R-module.

Modules are the basic tool of homological algebra. Examples of modules include the set of integers Z, the cubic lattice in d dimensions Zd, and the group ring of a group.

A module is one of the fundamental algebraic structures used in abstract algebra. A module over a ring is a generalization of the notion of vector space over a field, wherein the corresponding scalars are the elements of an arbitrary given ring (with identity) and a multiplication (on the left and/or on the right) is defined between elements of the ring and elements of the module. Thus, a module, like a vector space, is an additive abelian group; a product is defined between elements of the ring and elements of the module that is distributive over the addition operation of each parameter and is compatible with the ring multiplication.

Modules are very closely related to the representation theory of groups. They are also one of the central notions of commutative algebra and homological algebra, and are used widely in algebraic geometry and algebraic topology.

MOTIVATION

In a vector space, the set of scalars is a field and acts on the vectors by scalar multiplication, subject to certain axioms such as the distributive law. In a module, the scalars need only be a ring, so the module concept represents a significant generalization. In commutative algebra, both ideals and quotient rings are modules, so that many arguments about ideals or quotient rings can be combined into a single argument about modules. In non-commutative algebra the distinction between left ideals, ideals, and modules becomes more pronounced, though some ring-theoretic conditions can be expressed either about left ideals or left modules. Much of the theory of modules consists of extending as many of the desirable properties of vector spaces as possible to the realm of modules over a "well-behaved" ring, such as a principal ideal domain. However, modules can be quite a bit more complicated than vector spaces; for instance, not all modules have a basis, and even those that do, free modules, need not have a unique rank if the underlying ring does not satisfy the invariant basis number condition, unlike vector spaces, which always have a (possibly infinite) basis whose cardinality is then unique. (These last two assertions require the axiom of choice in general, but not in the case of finite-dimensional spaces, or certain well-behaved infinite-dimensional spaces such as Lp spaces.)

Formal definition

Suppose that R is a ring and 1R is its multiplicative identity. A left R-module M consists of an abelian group (M, +) and an operation. : R × M '! M such that for all r, s in R and x, y in M, we have:

$$r.\ (x + y) = r.x + r.y(r + s).\ x = r.x + s.x(rs).\ x = r.(s.x)$$

The operation of the ring on M is called scalar multiplication, and is usually written by juxtaposition, i.e. as rx for r in R and x in M, though here it is denoted as r ÅŠ x to distinguish it from the ring multiplication operation, denoted here by juxtaposition. The notation RM indicates a left R-module M. A right R-module M or MR is defined similarly, except that the ring acts on the right; i.e., scalar multiplication takes the form ÅŠ : M × R '! M, and the above axioms are written with scalars r and s on the right of x and y. Authors who do not require rings to be unital omit condition 4 above in the definition of an R-module, and so would call the structures defined above "unital left R-modules". In this article, consistent with the glossary of ring theory, all rings and modules are assumed to be unital.

If one writes the scalar action as fr so that fr(x) = r ÅŠ x, and f for the map that takes each r to its corresponding map fr , then the first axiom states that every fr is a group endomorphism of M, and the other three axioms assert that the map f : R

’! End(M) given by r ¦! fr is a ring homomorphism from R to the endomorphism ring End(M). Thus a module is a ring action on an abelian group (cf. group action. Also consider monoid action of multiplicative structure of R). In this sense, module theory generalizes representation theory, which deals with group actions on vector spaces, or equivalently group ring actions. A bimodule is a module that is a left module and a right module such that the two multiplications are compatible. If R is commutative, then left R-modules are the same as right R-modules and are simply called R-modules.

Examples

- If K is a field, then K-vector spaces (vector spaces over K) and K-modules are identical.
- If K is a field, and K[x] a univariate polynomial ring, then a K[x]-module M is a K-module with an additional action of x on M that commutes with the action of K on M. In other words, a K[x]-module is a K-vector space M combined with a linear map from M to M. Applying the Structure theorem for finitely generated modules over a principal ideal domain to this example shows the existence of the rational and Jordan canonical forms.
- The concept of a Z-module agrees with the notion of an abelian group. That is, every abelian group is a module over the ring of integers Z in a unique way. For n > 0, let n Å” x = x + x +... + x (n summands), 0 Å” x = 0, and (“n) Å” x = “(n Å” x). Such a module need not have a basis—groups containing torsion elements do not. (For example, in the group of integers modulo 3, one cannot find even one element which satisfies the definition of a linearly independent set since when an integer such as 3 or 6 multiplies an element, the result is 0. However, if a finite field is considered as a module over the same finite field taken as a ring, it is a vector space and does have a basis.)
- The decimal fractions (including negative ones) form a module over the integers. Only singletons are linearly independent sets, but there is no singleton that can serve as a basis, so the module has no basis and no rank.
- If R is any ring and n a natural number, then the cartesian product Rn is both a left and right R-module over if we use the component-wise operations. Hence when n = 1, R is an R-module, where the scalar multiplication is just ring multiplication. The case n = 0 yields the trivial R-module {0} consisting only of its identity element. Modules of this type are called free and if R has invariant basis number (e.g. any commutative ring or field) the number n is then the rank of the free module.

- If Mn(R) is the ring of n × n matrices over a ring R, M is an Mn(R)-module, and ei is the n × n matrix with 1 in the (i, i)-entry (and zeros elsewhere), then eiM is an R-module, since reim = eirm Š eiM. So M breaks up as the direct sum of R-modules, M = e1M •Š... •Š enM. Conversely, given an R-module M0, then M0•Šn is an Mn(R)-module. In fact, the category of R-modules and the category of Mn(R)-modules are equivalent. The special case is that the module M is just R as a module over itself, then Rn is an Mn(R)-module.
- If S is a nonempty set, M is a left R-module, and MS is the collection of all functions f : S '! M, then with addition and scalar multiplication in MS defined by (f + g)(s) = f(s) + g(s) and (rf)(s) = rf(s), MS is a left R-module. The right R-module case is analogous. In particular, if R is commutative then the collection of R-module homomorphisms h : M '! N (see below) is an R-module (and in fact a submodule of NM).
- If X is a smooth manifold, then the smooth functions from X to the real numbers form a ring C"(X). The set of all smooth vector fields defined on X form a module over C"(X), and so do the tensor fields and the differential forms on X. More generally, the sections of any vector bundle form a projective module over C"(X), and by Swan's theorem, every projective module is isomorphic to the module of sections of some bundle; the category of C"(X)-modules and the category of vector bundles over X are equivalent.
- If R is any ring and I is any left ideal in R, then I is a left R-module, and analogously right ideals in R are right R-modules.
- If R is a ring, we can define the ring Rop which has the same underlying set and the same addition operation, but the opposite multiplication: if ab = c in R, then ba = c in Rop. Any left R-module M can then be seen to be a right module over Rop, and any right module over R can be considered a left module over Rop.
- There are modules of a Lie algebra as well.

Submodules and homomorphisms

Suppose M is a left R-module and N is a subgroup of M. Then N is a submodule (or more explicitly an R-submodule) if for any n in N and any r in R, the product r Å" n is in N (or n Å" r for a right R-module). The set of submodules of a given module M, together with the two binary operations + and)", forms a lattice which satisfies the modular law: Given submodules U, N1, N2 of M such that N1 ," N2, then the following two submodules are equal: (N1 + U))" N2 = N1 + (U)" N2). If M and N are left R-modules, then a map f : M '! N is a homomorphism of R-modules if for any m, n in M and r, s in R,f(r·m + s·n) = r · f(m) + s · f (n)

This, like any homomorphism of mathematical objects, is just a mapping which preserves the structure of the objects. Another name for a homomorphism of R-modules is an R-linear map. A bijective module homomorphism is an isomorphism of modules, and the two modules are called isomorphic. Two isomorphic modules are identical for all practical purposes, differing solely in the notation for their elements.

The kernel of a module homomorphism f : M '! N is the submodule of M consisting of all elements that are sent to zero by f. The isomorphism theorems familiar from groups and vector spaces are also valid for R-modules. Given a ring R, the set of all left R-modules together with their module homomorphisms forms an abelian category, denoted by R-Mod (see category of modules).

Types of modules

Finitely generated. An R-module M is finitely generated if there exist finitely many elements x1,..., xn in M such that every element of M is a linear combination of those elements with coefficients from the ring R. Cyclic. A module is called a cyclic module if it is generated by one element. Free. A free R-module is a module that has a basis, or equivalently, one that is isomorphic to a direct sum of copies of the ring R. These are the modules that behave very much like vector spaces. Projective. Projective modules are direct summands of free modules and share many of their desirable properties. Injective. Injective modules are defined dually to projective modules. Flat. A module is called flat if taking the tensor product of it with any exact sequence of R-modules preserves exactness.

Torsionless module. A module is called torsionless if it embeds into its algebraic dual. Simple. A simple module S is a module that is not {0} and whose only submodules are {0} and S. Simple modules are sometimes called irreducible. Semisimple. A semisimple module is a direct sum (finite or not) of simple modules. Historically these modules are also called completely reducible. Indecomposable. An indecomposable module is a non-zero module that cannot be written as a direct sum of two non-zero submodules. Every simple module is indecomposable, but there are indecomposable modules which are not simple (e.g. uniform modules). Faithful. A faithful module M is one where the action of each r "" 0 in R on M is nontrivial (i.e. r Å" x "" 0 for some x in M). Equivalently, the annihilator of M is the zero ideal.

Torsion-free. A torsion-free module is a module over a ring such that 0 is the only element annihilated by a regular element (non zero-divisor) of the ring. Noetherian. A Noetherian module is a module which satisfies the ascending chain condition on submodules, that is, every increasing chain of submodules becomes stationary after finitely many steps.

Equivalently, every submodule is finitely generated. Artinian. An Artinian module is a module which satisfies the descending chain condition on submodules, that is, every decreasing chain of submodules becomes stationary after finitely many steps. Graded. A graded module is a module with a decomposition as a direct sum M = *x Mx over a graded ring R = *x Rx such that RxMy ," Mx+y for all x and y. Uniform. A uniform module is a module in which all pairs of nonzero submodules have nonzero intersection.

FURTHER NOTIONS

Relation to representation theory

If M is a left R-module, then the action of an element r in R is defined to be the map M '! M that sends each x to rx (or xr in the case of a right module), and is necessarily a group endomorphism of the abelian group (M, +). The set of all group endomorphisms of M is denoted EndZ(M) and forms a ring under addition and composition, and sending a ring element r of R to its action actually defines a ring homomorphism from R to EndZ(M).

Such a ring homomorphism R '! EndZ(M) is called a representation of R over the abelian group M; an alternative and equivalent way of defining left R-modules is to say that a left R-module is an abelian group M together with a representation of R over it. A representation is called faithful if and only if the map R '! EndZ(M) is injective. In terms of modules, this means that if r is an element of R such that rx = 0 for all x in M, then r = 0. Every abelian group is a faithful module over the integers or over some modular arithmetic Z/nZ.

Generalizations

Any ring R can be viewed as a preadditive category with a single object. With this understanding, a left R-module is just a covariant additive functor from R to the category Ab of abelian groups, and right R-modules are contravariant additive functors. This suggests that, if C is any preadditive category, a covariant additive functor from C to Ab should be considered a generalized left module over C. These functors form a functor category C-Mod which is the natural generalization of the module category R-Mod. Modules over commutative rings can be generalized in a different direction: take a ringed space (X, OX) and consider the sheaves of OX-modules (see sheaf of modules). These form a category OX-Mod, and play an important role in modern algebraic geometry. If X has only a single point, then this is a module category in the old sense over the commutative ring OX(X).

One can also consider modules over a semiring. Modules over rings are abelian groups, but modules over semirings are only commutative monoids. Most applications of modules are still possible. In particular, for any semiring S, the matrices over S form a semiring over which the tuples of elements from S are a module (in this generalized sense only). This allows a further generalization of the concept of vector space incorporating the semirings from theoretical computer science. Over near-rings, one can consider near-ring modules, a nonabelian generalization of modules.

4

BASICS OF SET THEORY

Set theory is the mathematical theory of well-determined collections, called sets, of objects that are called members, or elements, of the set. Pure set theory deals exclusively with sets, so the only sets under consideration are those whose members are also sets. The theory of the hereditarily-finite sets, namely those finite sets whose elements are also finite sets, the elements of which are also finite, and so on, is formally equivalent to arithmetic. So, the essence of set theory is the study of infinite sets, and therefore it can be defined as the mathematical theory of the actual—as opposed to potential—infinite.

The notion of set is so simple that it is usually introduced informally, and regarded as self-evident. In set theory, however, as is usual in mathematics, sets are given axiomatically, so their existence and basic properties are postulated by the appropriate formal axioms. The axioms of set theory imply the existence of a set-theoretic universe so rich that all mathematical objects can be construed as sets. Also, the formal language of pure set theory allows one to formalize all mathematical notions and arguments. Thus, set theory has become the standard foundation for mathematics, as every mathematical object can be viewed as a set, and every theorem of mathematics can be logically deduced in the Predicate Calculus from the axioms of set theory.

Both aspects of set theory, namely, as the mathematical science of the infinite, and as the foundation of mathematics, are of philosophical importance.

THE ORIGINS

Set theory, as a separate mathematical discipline, begins in the work of Georg Cantor. One might say that set theory was born in late 1873, when he made the amazing discovery that the linear continuum, that is, the real line, is not countable, meaning that its points cannot be counted using the natural numbers. So, even though the set of natural numbers and the set of real numbers are both infinite, there are more real numbers than there are natural numbers, which opened the door to the investigation of the different sizes of infinity. See the entry on the early development of set theory for a discussion of the origin of set-theoretic ideas and their use by different mathematicians and philosophers before and around Cantor's time.

According to Cantor, two sets A

and B have the same size, or cardinality, if they are bijectable, i.e., the elements of A can be put in a one-to-one correspondence with the elements of B. Thus, the set N of natural numbers and the set R of real numbers have different cardinalities. In 1878 Cantor formulated the famous Continuum Hypothesis (CH), which asserts that every infinite set of real numbers is either countable, i.e., it has the same cardinality as N, or has the same cardinality as R.

In other words, there are only two possible sizes of infinite sets of real numbers. The CH is the most famous problem of set theory. Cantor himself devoted much effort to it, and so did many other leading mathematicians of the first half of the twentieth century, such as Hilbert, who listed the CH as the first problem in his celebrated list of 23 unsolved mathematical problems presented in 1900 at the Second International Congress of Mathematicians, in Paris. The attempts to prove the CH led to major discoveries in set theory, such as the theory of constructible sets, and the forcing technique, which showed that the CH can neither be proved nor disproved from the usual axioms of set theory. To this day, the CH remains open.

Early on, some inconsistencies, or paradoxes, arose from a naive use of the notion of set; in particular, from the deceivingly natural assumption that every property determines a set, namely the set of objects that have the property. One example is Russell's Paradox, also known to Zermelo:

Consider the property of sets of not being members of themselves. If the property determines a set, call it A, then A is a member of itself if and only if A is not a member of itself.

Thus, some collections, like the collection of all sets, the collection of all ordinals numbers, or the collection of all cardinal numbers, are not sets. Such collections are called proper classes.

In order to avoid the paradoxes and put it on a firm footing, set theory had to be axiomatized. The first axiomatization was due to Zermelo (1908) and it came as a result of the need to spell out the basic set-theoretic principles underlying his proof of Cantor's Well-Ordering Principle. Zermelo's axiomatization avoids Russell's Paradox by means of the Separation axiom, which is formulated as quantifying over properties of sets, and thus it is a second-order statement. Further work by Skolem and Fraenkel led to the formalization of the Separation axiom in terms of formulas of first-order, instead of the informal notion of property, as well as to the introduction of the axiom of Replacement, which is also formulated as an axiom schema for first-order formulas (see next section). The axiom of Replacement is needed for a proper development of the theory of transfinite ordinals and cardinals, using transfinite recursion (see Section 3). It is also needed to prove the existence of such simple sets as the set of hereditarily finite sets, i.e., those finite sets whose elements are finite, the elements of which are also finite, and so on; or to prove basic set-theoretic facts such as that every set is contained in a transitive set, i.e., a set that contains all elements of its elements (see Mathias 2001 for the weaknesses of Zermelo set theory). A further addition, by von Neumann, of the axiom of Foundation, led to the standard axiom system of set theory, known as the Zermelo-Fraenkel axioms plus the Axiom of Choice, or ZFC.

Other axiomatizations of set theory, such as those of von Neumann-Bernays-Gödel (NBG), or Morse-Kelley (MK), allow also for a formal treatment of proper classes.

History of Set theory

Mathematical topics typically emerge and evolve through interactions among many researchers. Set theory, however, was founded by a single paper in 1874 by Georg Cantor: "On a Property of the Collection of All Real Algebraic Numbers".

Since the 5th century BC, beginning with Greek mathematician Zeno of Elea in the West and early Indian mathematicians in the East, mathematicians had struggled with the concept of infinity. Especially notable is the work of Bernard Bolzano in the first half of the 19th century. Modern understanding of infinity began in 1870–1874 and was motivated by Cantor's work in real analysis. An 1872 meeting between Cantor and Richard Dedekind influenced Cantor's thinking and culminated in Cantor's 1874 paper.

Cantor's work initially polarized the mathematicians of his day. While Karl Weierstrass and Dedekind supported Cantor, Leopold Kronecker, now seen as a founder of mathematical constructivism, did not. Cantorian set theory eventually became widespread, due to the utility of Cantorian concepts, such as one-to-one correspondence among sets, his proof that there are more real numbers than integers, and the "infinity of infinities" ("Cantor's paradise") resulting from the power set operation. This utility of set theory led to the article "Mengenlehre" contributed in 1898 by Arthur Schoenflies to Klein's encyclopedia.

The next wave of excitement in set theory came around 1900, when it was discovered that some interpretations of Cantorian set theory gave rise to several contradictions, called antinomies or paradoxes. Bertrand Russell and Ernst Zermelo independently found the simplest and best known paradox, now called Russell's paradox: consider "the set of all sets that are not members of themselves", which leads to a contradiction since it must be a member of itself and not a member of itself. In 1899 Cantor had himself posed the question "What is the cardinal number of the set of all sets?", and obtained a related paradox. Russell used his paradox as a theme in his 1903 review of continental mathematics in his The Principles of Mathematics. In 1906 English readers gained the book Theory of Sets of Points by husband and wife William Henry Young and Grace Chisholm Young, published by Cambridge University Press.

The momentum of set theory was such that debate on the paradoxes did not lead to its abandonment. The work of Zermelo in 1908 and the work of Abraham Fraenkel and Thoralf Skolem in 1922 resulted in the set of axioms ZFC, which became the most commonly used set of axioms for set theory. The work of analysts such as Henri Lebesgue demonstrated the great mathematical utility of set theory, which has since become woven into the fabric of modern mathematics. Set theory is commonly used as a foundational system, although in some areas—such as algebraic geometry and algebraic topology—category theory is thought to be a preferred foundation.

The axioms of set theory

ZFC is an axiom system formulated in first-order logic with equality and with only one binary relation symbol ? for membership. Thus, we write A?B to express that A is a member of the set B. See the Supplement on Basic Set Theory for further details. See also the Supplement on Zermelo-Fraenkel Set Theory for a formalized version of the axioms and further comments. We state below the axioms of ZFC informally.

The axioms of ZFC

Extensionality: If two sets A and B have the same elements, then they are equal. Null Set: There exists a set, denoted by Ø and called the empty set, which has no elements.

Pair: Given any sets A and B, there exists a set, denoted by {A,B}, which contains A and B as its only elements. In particular, there exists the set {A} which has A as its only element.

Power Set: For every set A there exists a set, denoted by P(A) and called the power set of A, whose elements are all the subsets of A

Union: For every set A, there exists a set, denoted by ?A and called the union of A, whose elements are all the elements of the elements of A

Infinity: There exists an infinite set. In particular, there exists a set Z that contains Ø and such that if A?Z, then ?{A,{A}}?Z

Separation: For every set A and every given property, there is a set containing exactly the elements of A that have that property. A property is given by a formula f of the first-order language of set theory.

Thus, Separation is not a single axiom but an axiom schema, that is, an infinite list of axioms, one for each formula f

Replacement: For every given definable function with domain a set A, there is a set whose elements are all the values of the function.

Replacement is also an axiom schema, as definable functions are given by formulas.

Foundation: Every non-empty set A

contains an ?-minimal element, that is, an element such that no element of A belongs to it.

These are the axioms of Zermelo-Fraenkel set theory, or ZF. The axioms of Null Set and Pair follow from the other ZF axioms, so they may be omitted. Also, Replacement implies Separation.

Finally, there is the Axiom of Choice (AC):

Choice: For every set A of pairwise-disjoint non-empty sets, there exists a set that contains exactly one element from each set in A.

The AC was, for a long time, a controversial axiom. On the one hand, it is very useful and of wide use in mathematics. On the other hand, it has rather unintuitive consequences, such as the Banach-Tarski Paradox, which says that the unit ball can be partitioned into finitely-many pieces, which can then be rearranged to form two unit balls. The objections to the axiom arise from the fact that it asserts the existence of sets that cannot be explicitly defined. But Gödel's 1938 proof of its consistency, relative to the consistency of ZF, dispelled any suspicions left about it.

The Axiom of Choice is equivalent, modulo ZF, to the Well-ordering Principle, which asserts that every set can be well-ordered, i.e., it can be linearly ordered so that every non-empty subset has a minimal element.

Although not formally necessary, besides the symbol? one normally uses for convenience other auxiliary defined symbols. For example, A B expresses that A is a subset of B, i.e., every member of A is a member of B. Other symbols are used to denote sets obtained by performing basic operations, such as A B, which denotes the union of A and B, i.e., the set whose elements are those of A and B; or AnB, which denotes the intersection of A and B, i.e., the set whose elements are those common to A and B. The ordered pair (A,B) is defined as the set {{A},{A,B}}. Thus, two ordered pairs (A,B) and (C,D) are equal if and only if A=C and B=D. And the Cartesian product A×B is defined as the set of all ordered pairs (C,D) such that C A and D B. Given any formula f(x,y1,...,yn), and sets A,B1,...,Bn, one can form the set of all those elements of A that satisfy the formula f(x,B1,...,Bn). This set is denoted by {a A:f(a,B1,...,Bn)}. In ZF one can easily prove that all these sets exist. See the Supplement on Basic Set Theory for further discussion.

Basic Concepts and Notation

Set theory begins with a fundamental binary relation between an object o and a set A. If o is a member (or element) of A, the notation o"A is used. A set is described by listing elements separated by commas, or by a characterizing property of its elements, within braces {}. Since sets are objects, the membership relation can relate sets as well. A derived binary relation between two sets is the subset relation, also called set inclusion. If all the members of set A are also members of set B, then A is a subset of B, denoted

A†"B. For example, {1, 2} is a subset of {1, 2, 3}, and so is {2} but {1, 4} is not. As insinuated from this definition, a set is a subset of itself. For cases where this possibility is unsuitable or would make sense to be rejected, the term proper subset is defined. A is called a proper subset of B if and only if A is a subset of B, but A is not equal to B. Note also that 1, 2, and 3 are members (elements) of the set {1, 2, 3} but are not subsets of it; and in turn, the subsets, such as {1}, are not members of the set {1, 2, 3}.

Just as arithmetic features binary operations on numbers, set theory features binary operations on sets. The:

- Union of the sets A and B, denoted A*"B, is the set of all objects that are a member of A, or B, or both. The union of {1, 2, 3} and {2, 3, 4} is the set {1, 2, 3, 4}.

- Intersection of the sets A and B, denoted A) B, is the set of all objects that are members of both A and B. The intersection of {1, 2, 3} and {2, 3, 4} is the set {2, 3}.
- Set difference of U and A, denoted U \ A, is the set of all members of U that are not members of A. The set difference {1, 2, 3} \ {2, 3, 4} is {1}, while, conversely, the set difference {2, 3, 4} \ {1, 2, 3} is {4}. When A is a subset of U, the set difference U \ A is also called the complement of A in U. In this case, if the choice of U is clear from the context, the notation Ac is sometimes used instead of U \ A, particularly if U is a universal set as in the study of Venn diagrams.
- Symmetric difference of sets A and B, denoted A³%B or A–"B, is the set of all objects that are a member of exactly one of A and B (elements which are in one of the sets, but not in both). For instance, for the sets {1, 2, 3} and {2, 3, 4}, the symmetric difference set is {1, 4}. It is the set difference of the union and the intersection, (A*"B) \ (A)" B) or (A \ B) *" (B \ A).
- Cartesian product of A and B, denoted A × B, is the set whose members are all possible ordered pairs (a, b) where a is a member of A and b is a member of B. The cartesian product of {1, 2} and {red, white} is {(1, red), (1, white), (2, red), (2, white)}.
- Cartesian product of A and B, denoted A × B, is the set whose members are all possible ordered pairs (a, b) where a is a member of A and b is a member of B. The cartesian product of {1, 2} and {red, white} is {(1, red), (1, white), (2, red), (2, white)}.

Some basic sets of central importance are the empty set (the unique set containing no elements; occasionally called the null set though this name is ambiguous), the set of natural numbers, and the set of real numbers.

Algebra of Sets

The algebra of sets defines the properties and laws of sets, the set-theoretic operations of union, intersection, and complementation and the relations of set equality and set inclusion. It also provides systematic procedures for evaluating expressions, and performing calculations, involving these operations and relations. Any set of sets closed under the set-theoretic operations forms a Boolean algebra with the join operator being union, the meet operator being intersection, the complement operator being set complement, the bottom being and the top being the universe set under consideration.

Fundamentals

The algebra of sets is the set-theoretic analogue of the algebra of numbers. Just as arithmetic addition and multiplication are associative and commutative, so are set union and intersection; just as the arithmetic relation"less than or equal" is reflexive, antisymmetric and transitive, so is the set relation of "subset". It is the algebra of the set-theoretic operations of union, intersection and complementation, and the relations of equality and inclusion. For a basic introduction to sets see the article on sets, for a fuller account see naive set theory, and for a full rigorous axiomatic treatment see axiomatic set theory.

The Fundamental laws of Set Algebra

The binary operations of set union () and intersection () satisfy many identities. Several of these identities or "laws" have well established names.

Commutative laws:

- Associative laws:
- Distributive laws:
- The union and intersection of sets may be seen as analogous to the addition and multiplication of numbers. Like addition and multiplication, the operations of union and intersection are commutative and associative, and intersection distributes over union. However, unlike addition and multiplication, union also distributes over intersection.

Two additional pairs of laws involve the special sets called the empty set Ø and the universe set ; together with the complement operator (denotes the complement of. This can also be written as, read as A prime). The empty set has no members, and the universe set has all possible members (in a particular context).

Identity laws:

- Complement laws:
- The identity laws (together with the commutative laws) say that, just like 0 and 1 for addition and multiplication, Ø and U are the identity elements for union and intersection, respectively.

Unlike addition and multiplication, union and intersection do not have inverse elements. However the complement laws give the fundamental properties of the somewhat inverse-like unary operation of set complementation. The preceding five pairs of laws-the commutative, associative, distributive, identity and complement laws-encompass all of set algebra, in the sense that every valid proposition in the algebra of sets can be derived from them.

Note that if the complement laws are weakened to the rule, then this is exactly the algebra of propositional linear logic.

The principle of duality

Each of the identities stated above is one of a pair of identities such that each can be transformed into the other by interchanging *" and)", and also Ø and U. These are examples of an extremely important and powerful property of set algebra, namely, the principle of duality for sets, which asserts that for any true statement about sets, the dual statement obtained by interchanging unions and intersections, interchanging U and Ø and reversing inclusions is also true. A statement is said to be self-dual if it is equal to its own dual.

Some additional laws for unions and intersections The following proposition states six more important laws of set algebra, involving unions and intersections. PROPOSITION 3: For any subsets A and B of a universe set U, the following identities hold:

Idempotent laws:

- Domination laws:
- Absorption laws:
- As noted above, each of the laws stated in proposition 3 can be derived from the five fundamental pairs of laws stated above. As an illustration, a proof is given below for the idempotent law for union.

Proof:

		by the identity law of intersection
		by the complement law for union
		by the distributive law of union over intersection
		by the complement law for intersection
		by the identity law for union

The following proof illustrates that the dual of the above proof is the proof of the dual of the idempotent law for union, namely the idempotent law for intersection.

Proof:

		by the identity law for union
		by the complement law for intersection
		by the distributive law of intersection over union
		by the complement law for union
		by the identity law for intersection

Intersection can be expressed in terms of set difference :

Some additional laws for complements

The following proposition states five more important laws of set algebra, involving complements.

PROPOSITION 4: Let A and B be subsets of a universe U, then:

De Morgan's laws:

- Double complement or involution law:
- Complement laws for the universe set and the empty set:
- Notice that the double complement law is self-dual.

The next proposition, which is also self-dual, says that the complement of a set is the only set that satisfies the complement laws. In other words, complementation is characterized by the complement laws.

PROPOSITION 5: Let A and B be subsets of a universe U, then:

uniqueness of complements:

- If, and, then

The Algebra of Inclusion

The following proposition says that inclusion, that is the binary relation of one set being a subset of another, is a partial order.

PROPOSITION 6: If A, B and C are sets then the following hold:

reflexivity:

- Antisymmetry:
- And if and only if

transitivity:

- If and, then

The following proposition says that for any set S, the power set of S, ordered by inclusion, is a bounded lattice, and hence together with the distributive and complement laws above, show that it is a Boolean algebra.

PROPOSITION 7: If A, B and C are subsets of a set S then the following hold:

existence of a least element and a greatest element:

- Existence of joins:
- If and, then

The following proposition says that the statement is equivalent to various other statements involving unions, intersections and complements.

PROPOSITION 8: For any two sets A and B, the following are equivalent:

- The above proposition shows that the relation of set inclusion can be characterized by either of the operations of set union or set intersection, which means that the notion of set inclusion is axiomatically superfluous.

The Algebra of Relative Complements

The following proposition lists several identities concerning relative complements and set-theoretic differences.

PROPOSITION 9: For any universe U and subsets A, B, and C of U, the following identities hold:

- The above proposition shows that the relation of set inclusion can be characterized by either of the operations of set union or set intersection, which means that the notion of set inclusion is axiomatically superfluous.

The Algebra of Relative Complements

The following proposition lists several identities concerning relative complements and set-theoretic differences.

PROPOSITION 9: For any universe U and subsets A, B, and C of U, the following identities hold:

DUALITY (ORDER THEORY)

In the mathematical area of order theory, every partially ordered set P gives rise to a dual (or opposite) partially ordered set which is often denoted by Pop or Pd. This dual order Pop is defined to be the set with the inverse order, i.e. x d" y holds in Pop if and only if y d" x holds in P. It is easy to see that this construction, which can be depicted by flipping the Hasse diagram for P upside down, will indeed yield a partially ordered set. In a broader sense, two posets are also said to be duals if they are dually isomorphic, i.e. if one poset is order isomorphic to the dual of the other.

The importance of this simple definition stems from the fact that every definition and theorem of order theory can readily be transferred to the dual order. Formally, this is captured by the Duality Principle for ordered sets: If a given statement is valid for all partially ordered sets, then its dual statement, obtained by inverting the direction of all order relations and by dualizing all order theoretic definitions involved, is also valid for all partially ordered sets.If a statement or definition is equivalent to its dual then it is said to be self-dual. Note that the consideration of dual orders is so fundamental that it often occurs implicitly when writing e" for the dual order of d" without giving any prior definition of this "new" symbol.

Examples

A Bounded Distributive lattice, and its Dual

Naturally, there are a great number of examples for concepts that are dual:

- Greatest elements and least elements
- Maximal elements and minimal elements
- Least upper bounds (suprema, (") and greatest lower bounds
- (infima, "')
- Upper sets and lower sets
- Ideals and filters
- Closure operators and kernel operators.

Examples of notions which are self-dual include:

- Being a (complete) lattice
- Monotonicity of functions
- Distributivity of lattices, i.e. the lattices for which "x,y,z: x "' (y (" z) = (x "' y) (" (x "' z) holds are exactly those for which the dual
- Statement "x,y,z: x (" (y "' z) = (x (" y) "' (x (" z) holds
- Being a Boolean algebra
- Being an order isomorphism.

Since partial orders are antisymmetric, the only ones that are self-dual are the equivalence relations.

Duality (mathematics)

In mathematics, a duality, generally speaking, translates concepts, theorems or mathematical structures into other concepts, theorems or structures, in a one-to-one fashion, often (but not always) by means of an involution operation: if the dual of A is B, then the dual of B is A. Such involutions sometimes have fixed points, so that the dual of A is A itself. For example, Desargues' theorem is self-dual in this sense under the standard duality in projective geometry. In mathematical contexts, duality has numerous meanings although it is "a very pervasive and important concept in (modern) mathematics" and "an important general theme that has manifestations in almost every area of mathematics".

Many mathematical dualities between objects of two types correspond to pairings, bilinear functions from an object of one type and another object of the second type to some family of scalars. For instance, linear algebra duality corresponds in this way to bilinear maps from pairs of vector spaces to scalars, the duality between distributions and the associated test functions corresponds to the pairing in which one integrates a distribution against a test function, and Poincaré duality corresponds similarly to intersection number, viewed as a pairing between submanifolds of a given manifold.

From a category theory viewpoint, duality can also be seen as a functor, at least in the realm of vector spaces. This functor assigns to each space its dual space, and the pullback construction assigns to each arrow f: V '! W its dual f": W" '! V".

Introductory Examples

In the words of Michael Atiyah,

Duality in mathematics is not a theorem, but a "principle". The following list of examples shows the common features of many dualities, but also indicates that the precise meaning of duality may vary from case to case.

Complement of a Subset

A simple, maybe the most simple, duality arises from considering subsets of a fixed set S. To any subset A †" S, the complement A consists of all those elements in S which are not contained in A. It is again a subset of S. Taking the complement has the following properties:

- Applying it twice gives back the original set, i.e., (Ac)c = A. This is referred to by saying that the operation of taking the complement is an involution.
- An inclusion of sets A †" B is turned into an inclusion in the opposite direction Bc †" Ac.
- Given two subsets A and B of S, A is contained in Bc if and only if B is contained in Ac.

This duality appears in topology as a duality between open and closed subsets of some fixed topological space X: a subset U of X is closed if and only if its complement in X is open. Because of this, many theorems about closed sets are dual to theorems about open sets. For example, any union of open sets is open, so dually, any intersection of closed sets is closed. The interior of a set is the largest open set contained in it, and the closure of the set is the smallest closed set that contains it. Because of the duality, the complement of the interior of any set U is equal to the closure of the complement of U.

Dual cone

A set C (blue) and its Dual cone C* (red).

A duality in geometry is provided by the dual cone construction. Given a set of points in the plane (or more generally points in), the dual cone is defined as the set consisting of those points satisfying

for all points in, as illustrated in the diagram. Unlike for the complement of sets mentioned above, it is not in general true that applying the dual cone construction twice gives back the original set. Instead, is the smallest cone containing which may be bigger than. Therefore this duality is weaker than the one above, in that

- Applying the operation twice gives back a possibly bigger set: for all, is contained in. (For some, namely the cones, the two are actually equal.)

The other two properties carry over without change:

- It is still true that an inclusion is turned into an inclusion in the opposite direction ().
- Given two subsets and of the plane, is contained in if and only if is contained in.

Dual Vector Space

A very important example of a duality arises in linear algebra by associating to any vector space V its dual vector space V*. Its elements are the k-linear maps, where k is the field over which V is defined. The three properties of the dual cone carry over to this type of duality by replacing subsets of by vector space and inclusions of such subsets by linear maps. That is:

- Applying the operation of taking the dual vector space twice gives another vector space V**. There is always a map V→V**. For some V, namely precisely the finite-dimensional vector spaces, this map is an isomorphism.
- A linear map V '! W gives rise to a map in the opposite direction (W* -V*).
- Given two vector spaces V and W, the maps from V to W* correspond to the maps from W to V*.

A particular feature of this duality is that V and V* are isomorphic for certain objects, namely finite-dimensional vector spaces. However, this is in a sense a lucky coincidence, for giving such an isomorphism requires a certain choice, for example the choice of a basis of V. This is also true in the case if V is a Hilbert space, via the Riesz representation theorem.

Galois Theory

In all the dualities discussed before, the dual of an object is of the same kind as the object itself. For example, the dual of a vector space is again a vector space. Many duality statements are not of this kind. Instead, such dualities reveal a close relation between objects of seemingly different nature. One example of such a more general duality is from Galois theory For a fixed Galois extension K / F, one may associate the Galois group Gal(K/E) to any intermediate field E(i.e., F ⊆ E ⊆ K). This group is a subgroup of the Galois group G = Gal(K/F). Conversely, to any such subgroup H ⊆ G there is the fixed field KH consisting of elements fixed by the elements in H.

Compared to the above, this duality has the following features:

An extension F †F2 of intermediate fields gives rise to an inclusion of Galois groups in the opposite direction: Gal(K/F2) †

- Gal(K/F).
- Associating Gal(K/E) to E and KH to H are inverse to each other. This is the content of the fundamental theorem of Galois theory.

Order-reversing dualities

Hasse diagram of the power set of {1,2,3,4}, partially ordered by ,". The dual poset, i.e. ordering by *f*", is obtained by turning the diagram upside-down. The green nodes form an upper set and a lower set in the original and the dual order, respectively. Given a poset P = (X, d") (short for partially ordered set; i.e., a set that has a notion of ordering but in which two elements cannot necessarily be placed in order relative to each other), the dual poset Pd = (X, e") comprises the same ground set but the converse relation. Familiar examples of dual partial orders include

- The subset and superset relations ," and *f*" on any collection of sets, such as the subsets of a fixed set S. This gives rise to the first example of a duality mentioned above.
- The divides and multiple-of relations on the integers.
- The descendant-of and ancestor-of relations on the set of humans.

A duality transform is an involutive antiautomorphism f of a partially ordered set S, that is, an order-reversing involution f : S '! S. In several important cases these simple properties determine the transform uniquely up to some simple symmetries. For example, if f1, f2 are two duality transforms then their composition is an order automorphism of S; thus, any two duality transforms differ only by an order automorphism. For example, all order automorphisms of a power set S =

2R are induced by permutations of R. A concept defined for a partial order P will correspond to a dual concept on the dual poset Pd. For instance, a minimal element of P will be a maximal element of Pd: minimality and maximality are dual concepts in order theory. Other pairs of dual concepts are upper and lower bounds, lower sets and upper sets, and ideals and filters.

In topology, open sets and closed sets are dual concepts: the complement of an open set is closed, and vice versa. In matroid theory, the family of sets complementary to the independent sets of a given matroid themselves form another matroid, called the dual matroid.

DIMENSION-REVERSING DUALITIES

The features of the cube and its dual octahedron correspond one-for-one with dimensions reversed.with a reversal of the dimensions of the features of the objects. A classical example of this is the duality of the platonic solids, in which the cube and the octahedron form a dual pair, the dodecahedron and the icosahedron form a dual pair, and the tetrahedron is self-dual. The dual polyhedron of any of these polyhedra may be formed as the convex hull of the center points of each face of the primal polyhedron, so the vertices of the dual correspond one-for-one with the faces of the primal. Similarly, each edge of the dual corresponds to an edge of the primal, and each face of the dual corresponds to a vertex of the primal. These correspondences are incidence-preserving: if two parts of the primal polyhedron touch each other, so do the corresponding two parts of the dual polyhedron.

More generally, using the concept of polar reciprocation, any convex polyhedron, or more generally any convex polytope, corresponds to a dual polyhedron or dual polytope, with an i-dimensional feature of an n-dimensional polytope corresponding to an (n " i " 1)-dimensional feature of the dual polytope. The incidence-preserving nature of the duality is reflected in the fact that the face lattices of the primal and dual polyhedra or polytopes are themselves order-theoretic duals. Duality of polytopes and order-theoretic duality are both involutions: the dual polytope of the dual polytope of any polytope is the original polytope, and reversing all order-relations twice returns to the original order. Choosing a different center of polarity leads to geometrically different dual polytopes, but all have the same combinatorial structure.

A planar Graph in Blue, and its Dual Graph in Red.

From any three-dimensional polyhedron, one can form a planar graph, the graph of its vertices and edges. The dual polyhedron has a dual graph, a graph with one vertex for each face of the polyhedron and with one edge for every two adjacent faces. The same concept of planar graph duality may be generalized to graphs that are drawn in the plane but that do not come from a three-dimensional polyhedron, or more generally to graph embeddings on surfaces of higher genus: one may draw a dual graph by placing one vertex within each region bounded by a cycle of edges in the embedding, and drawing an edge connecting any two regions that share a boundary edge.

An important example of this type comes from computational geometry: the duality for any finite set S of points in the plane between the Delaunay triangulation of S and the Voronoi diagram of S.

As with dual polyhedra and dual polytopes, the duality of graphs on surfaces is a dimension-reversing involution: each vertex in the primal embedded graph corresponds to a region of the dual embedding, each edge in the primal is crossed by an edge in the dual, and each region of the primal corresponds to a vertex of the dual.

The dual graph depends on how the primal graph is embedded: different planar embeddings of a single graph may lead to different dual graphs. Matroid duality is an algebraic extension of planar graph duality, in the sense that the dual matroid of the graphic matroid of a planar graph is isomorphic to the graphic matroid of the dual graph.

A kind of geometric duality also occurs in optimization theory, but not one that reverses dimensions. A linear program may be specified by a system of real variables (the coordinates for a point in Euclidean space), a system of linear constraints (specifying that the point lie in a halfspace; the intersection of these halfspaces is a convex polytope, the feasible region of the program), and a linear function (what to optimize). Every linear program has a dual problem with the same optimal solution, but the variables in the dual problem correspond to constraints in the primal problem and vice versa.

Duality in logic and Set Theory

In logic, functions or relations A and B are considered dual if A(¬x) = ¬B(x), where ¬ is logical negation. The basic duality of this type is the duality of the “ and “ quantifiers in classical logic. These are dual because “x.¬P(x) and ¬”x.P(x) are equivalent for all predicates P in classical logic: if there exists an x for which P fails to hold, then it is false that P holds for all x (but the converse does not hold constructively). From this fundamental logical duality follow several others:

- A formula is said to be satisfiable in a certain model if there are assignments to its free variables that render it true; it is valid if every assignment to its free variables makes it true. Satisfiability and validity are dual because the invalid formulas are precisely those whose negations are satisfiable, and the unsatisfiable formulas are those whose negations are valid. This can be viewed as a special case of the previous item, with the quantifiers ranging over interpretations.
- In classical logic, the "" and (" operators are dual in this sense, because (¬x "" ¬y) and ¬(x (" y) are equivalent. This means that for
- Every theorem of classical logic there is an equivalent dual theorem. De Morgan's laws are examples. More generally, "" (¬ xi) = ¬(" xi. The left side is true if and only if "i.¬xi, and the right side if and only if ¬"i.xi.
- In modal logic, ¡% p means that the proposition p is "necessarily" true, and Ê% p that p is "possibly" true. Most interpretations of modal logic assign dual meanings to these two operators. For example in Kripke semantics, "p is possibly true" means "there exists some world W such that p is true in W", while "p is necessarily true" means "for all worlds W, p is true in W". The duality of ¡% and Ê% then follows from the analogous duality of " and ". Other dual modal operators behave similarly. For example, temporal logic has operators denoting "will be true at some time in the future" and "will be true at all times in the future" which are similarly dual.

Other analogous dualities follow from these:

- Set-theoretic union and intersection are dual under the set complement operator Å"C. That is, AC)" BC = (A *" B)C, and more generally,)" AC

Dual objects

A group of dualities can be described by endowing, for any mathematical object X, the set of morphisms Hom (X, D) into some fixed object D, with a structure similar to that of X. This is sometimes called internal Hom. In general, this yields a true duality only for specific choices of D, in which case X* = Hom (X, D) is referred to as the dual of X. There is always a map from X to the bidual, that is to say, the dual of the dual,

It assigns to some x " X the map that associates to any map f : X '! D (i.e., an element in Hom(X, D)) the value f(x). Depending on the concrete duality considered and also depending on the object X, this map may or may not be an isomorphism.

Dual Vector Spaces Revisited

The Construction of the Dual Vector Space

Mentioned in the introduction is an example of such a duality. Indeed, the set of morphisms, i.e., linear maps, forms a vector space in its own right. The map V '! V** mentioned above is always injective. It is surjective, and therefore an isomorphism, if and only if the dimension of V is finite. This fact characterizes finite-dimensional vector spaces without referring to a basis.

ISOMORPHISMS OF V AND V" AND INNER PRODUCT SPACES

A vector space V is isomorphic to V" precisely if V is finite-dimensional. In this case, such an isomorphism is equivalent to a non-degenerate bilinear form In this case V is called an inner product space. For example, if K is the field of real or complex numbers, any positive definite bilinear form gives rise to such an isomorphism. In Riemannian geometry, V is taken to be the tangent space of a manifold and such positive bilinear forms are called Riemannian metrics. Their purpose is to measure angles and distances. Thus, duality is a foundational basis of this branch of geometry. Another application of inner product spaces is the Hodge star which provides a correspondence between the elements of the exterior algebra. For an n-dimensional vector space, the Hodge star operator maps forms to (n k)-forms. This can be used to formulate Maxwell's equations. In this guise, the duality inherent in the inner product space exchanges the role of magnetic and electric fields.

Duality in Projective Geometry

The complete quadrangle, a configuration of four points and six lines in the projective plane (left) and its dual configuration, the complete quadrilateral, with four lines and six points (right).

In some projective planes, it is possible to find geometric transformations that map each point of the projective plane to a line, and each line of the projective plane to a point, in an incidence-preserving way. For such planes there arises a general principle of duality in projective planes: given any theorem in such a plane projective geometry, exchanging the terms "point" and "line" everywhere results in a new, equally valid theorem. A simple example is that the statement "two points determine a unique line, the line passing through these points" has the dual statement that "two lines determine a unique point, the intersection point of these two lines". For further examples, see Dual theorems.

A conceptual explanation of this phenomenon in some planes (notably field planes) is offered by the dual vector space. In fact, the points in the projective plane correspond to one-dimensional subvector spaces while the lines in the projective plane correspond to subvector spaces of dimension 2. The duality in such projective geometries stems from assigning to a one-dimensional the subspace of consisting of those linear maps which satisfy. As a consequence of the dimension formula of linear algebra, this space is two-dimensional, i.e., it corresponds to a line in the projective plane associated to.

The (Positive Definite) Bilinear form

yields an identification of this projective plane with the. Concretely, the duality assigns to its orthogonal. The explicit formulas in duality in projective geometry arise by means of this identification.

Topological Vector Spaces and Hilbert Spaces

In the realm of topological vector spaces, a similar construction exists, replacing the dual by the topological dual vector space. A topological vector space that is canonically isomorphic to its bidual is called reflexive space. Forexample, the dual of an Lp-space is Lq where $1/p + 1/q = 1$ provided that $1 \leq p < \infty$, but the dual of L^∞

is bigger than L1 Hence L1 is not reflexive. Hilbert spaces H are equipped with an inner product (-,-) As in the finite-dimensional case, it allows to define a map The Riesz representation theorem asserts that this map is an isomorphism, i.e., every Hilbert space is isomorphic to its dual. Distributions are linear functionals on appropriate spaces of functions. They are an important technical means in the theory of partial differential equations (PDE): instead of solving a PDE directly, it may be easier to first solve the PDE in the "weak sense", i.e., find a distribution that satisfies the PDE and, second, to show that the solution must in fact be a function.

FURTHER DUAL OBJECTS

The dual lattice of a lattice L is Given by

which is used in the construction of toric varieties. The Pontryagin dual of locally compact topological groups G is given by continuous group homomorphisms with values in the circle (with multiplication of complex numbers as group operation).

Dual Categories

Opposite Category and Adjoint Functors

In another group of dualities, the objects of one theory are translated into objects of another theory and the maps between objects in the first theory are translated into morphisms in the second theory, but with direction reversed. Using the parlance of category theory, this amounts to a contravariant functor between two categories C and D: F: C- D which for any two objects X and Y of C gives a map HomC(X, Y) '! HomD(F(Y), F(X))

That functor may or may not be an equivalence of categories. There are various situations, where such a functor is an equivalence between the opposite category Cop of C, and D. Using a duality of this type, every statement in the first theory can be translated into a "dual" statement in the second theory, where the direction of all arrows has to be reversed. Therefore, any duality between categories C and D is formally the same as an equivalence between C and Dop (Cop and D). However, in many circumstances the opposite categories have no inherent meaning, which makes duality an additional, separate concept.

A category that is equivalent to its dual is called self-dual. An example of self-dual category is the category of Hilbert spaces.

Many category-theoretic notions come in pairs in the sense that they correspond to each other while considering the opposite category. For

example, Cartesian products Y1 × Y2 and disjoint unions Y1 """ Y2 of sets are dual to each other in the sense that

Hom (X, Y1 × Y2) = Hom (X, Y1) × Hom (X, Y2)

and

Hom (Y1 """ Y2, X) = Hom (Y1, X) × Hom (Y2, X)

for any set X. This is a particular case of a more general duality phenomenon, under which limits in a category C correspond to colimits in the opposite category Cop; further concrete examples of this are epimorphisms vs. monomorphism, in particular factor modules (or groups etc.) vs. submodules, direct products vs. direct sums (also called coproducts to emphasize the duality aspect). Therefore, in some cases, proofs of certain statements can be halved, using such a duality phenomenon. Further notions displaying related by such a categorical duality are projective and injective modules in homological algebra, fibrations and cofibrations in topology and more generally model categories.

Two functors F: C '! D and G: D '! C are adjoint if for all objects c in C and d in D HomD(F(c), d) E" HomC(c, G(d)),

in a natural way. Actually, the correspondence of limits and colimits is an example of adjoints, since there is an adjunction

colim: CI ? C: ?

between the colimit functor that assigns to any diagram in C indexed by some category I its colimit and the diagonal functor that maps any object c of C to the constant diagram which has c at all places. Dually,

?: C ? CI: lim.

Spaces and Functions

Gelfand duality is a duality between commutative C*-algebras A and compact Hausdorff spaces X is the same: it assigns to X the space of continuous functions (which vanish at infinity) from X to C, the complex numbers. Conversely, the space X can be reconstructed from A as the spectrum of A. Both Gelfand and Pontryagin duality can be deduced in a largely formal, category-theoretic way. In a similar vein there is a duality in algebraic geometry between commutative rings and affine schemes: to every commutative ring A there is an affine spectrum, Spec HYPERLINK "https://en.wikipedia.org/wiki/Spectrum_of_a_ring"A. Conversely, given an affine scheme S, one gets back a ring by taking global sections of the structure sheaf OS. In addition, ring homomorphisms are in one-to-one correspondence with morphisms of affine schemes, thereby there is an equivalence

(Commutative Rings) opE" (affine schemes)

Affine schemes are the local building blocks of schemes. The previous result therefore tells that the local theory of schemes is the same as commutative algebra, the study of commutative rings.

Noncommutative geometry draws inspiration from Gelfand duality and studies noncommutative C*-algebras as if they were functions on some imagined space. Tannaka–Krein duality is a non-commutative analogue of Pontryagin duality.

Galois Connections

In a number of situations, the two categories which are dual to each other are actually arising from partially ordered sets, i.e., there is some notion of an object "being smaller" than another one. A duality that respects the orderings in question is known as a Galois connection. An example is the standard duality in Galois theory mentioned in the introduction: a bigger field extension corresponds—under the mapping that assigns to any extension L ⊃ K (inside some fixed bigger field Ω) the Galois group Gal (Ω/ L) —to a smaller group.

The collection of all open subsets of a topological space X forms a complete Heyting algebra. There is a duality, known as Stone duality, connecting sober spaces and spatial locales.

- Birkhoff's representation theorem relating distributive lattices and partial orders

Pontryagin duality

Pontryagin duality gives a duality on the category of locally compact abelian groups: given any such group G, the character group x(G) Hom (G, S^1)

given by continuous group homomorphisms from G to the circle group S1 can be endowed with the compact-open topology. Pontryagin duality states that the character group is again locally compact abelian and that G ≅ x (x(G))

Moreover, discrete groups correspond to compact abelian groups; finite groups correspond to finite groups. On the one hand, Pontryagin is a special case of Gelfand duality. On the other hand, it is the conceptual reason of Fourier analysis, see below.

Analytic Dualities

In analysis, problems are frequently solved by passing to the dual description of functions and operators. Fourier transform switches between functions on a vector space and its dual:

and conversely If f is an L2-function on R or RN, say, then so is and Moreover, the transform interchanges operations of multiplication and convolution on the corresponding function spaces. A conceptual explanation of the Fourier transform is obtained by the aforementioned Pontryagin duality, applied to the locally compact groups R (or RN etc.): any character of R is given by $\xi \mapsto e^{-2\pi i x \xi}$ The dualizing character of Fourier transform has many other manifestations, for example, in alternative descriptions of quantum mechanical systems in terms of coordinate and momentum representations.

- Laplace transform is similar to Fourier transform and interchanges operators of multiplication by polynomials with constant coefficient linear differential operators.
- Legendre transformation is an important analytic duality which switches between velocities in Lagrangian mechanics and momenta in Hamiltonian mechanics.

Homology and cohomology

Theorems showing that certain objects of interest are the dual spaces (in the sense of linear algebra) of other objects of interest are often called dualities. Many of these dualities are given by a bilinear pairing of two K-vector spaces A⊗B→K

For perfect pairings, there is, therefore, an isomorphism of A to the dual of B.

Poincaré Duality

Poincaré duality of a smooth compact complex manifold X is given by a pairing of singular cohomology with C-coefficients (equivalently, sheaf cohomology of the constant sheaf C) H^i (X)⊗$H2^{n-i}$ (X)→C where n is the (complex) dimension of X. Poincaré duality can also be expressed as a relation of singular homology and de Rham cohomology, by asserting that the map (integrating a differential k-form over an 2n"k-(real) -dimensional cycle) is a perfect pairing. Poincaré duality also reverses dimensions; it corresponds to the fact that, if a topological manifold is represented as a cell complex, then the dual of the complex (a higher-dimensional generalization of the planar graph dual) represents the same manifold. In Poincaré duality, this homeomorphism is reflected in an isomorphism of the kth homology group and the (n " k)th cohomology group.

DUALITY IN ALGEBRAIC AND ARITHMETIC GEOMETRY

The same duality pattern holds for a smooth projective variety over a separably closed field, using l-adic cohomology with Q–!-coefficients instead. This is further generalized to possibly singular varieties, using intersection cohomology instead, a duality called Verdier duality. Serre duality or coherent duality are similar to the statements above, but applies to cohomology of coherent sheaves instead. With increasing level of generality, it turns out, an increasing amount of technical background is helpful or necessary to understand these theorems: the modern formulation of these dualities can be done using derived categories and certain direct and inverse image functors of sheaves (with respect to the classical analytical topology on manifolds for Poincaré duality, l-adic sheaves and the étale topology in the second case, and with respect to coherent sheaves for coherent duality).

Yet another group of similar duality statements is encountered in arithmetics: étale cohomology of finite, local and global fields (also known as Galois cohomology, since étale cohomology over a field is equivalent to group cohomology of the (absolute) Galois group of the field) admit similar pairings. The absolute Galois group G(Fq) of a finite field, for example, is isomorphic to the profinite completion of Z, the integers. Therefore, the perfect pairing (for any G-module M)

$$H^n\left(G,M\right) \times H^{1-n}\left(G,\ Hom\ \left(M,\mathbf{Q}/\mathbf{Z}\right)\right) \to \mathbf{Q}/\mathbf{Z}$$

is a direct consequence of Pontryagin duality of finite groups. For local and global fields, similar statements exist (local duality and global or Poitou–Tate duality).

C " (A)" B) = (C " A) *" (C " B) (A\cap B)=(C\setminus A)\cup (C\setminus B)Some ontology

A set is pure if all of its members are sets, all members of its members are sets, and so on. For example, the set {{}} containing only the empty set is a nonempty pure set. In modern set theory, it is common to restrict attention to the von Neumann universe of pure sets, and many systems of axiomatic set theory are designed to axiomatize the pure sets only. There are many technical advantages to this restriction, and little generality is lost, because essentially all mathematical concepts can be modeled by pure sets. Sets in the von Neumann universe are organized into a cumulative hierarchy, based on how deeply their members, members of members, etc. are nested. Each set in this hierarchy is assigned (by transfinite recursion) anordinal number á, known as its rank. The rank of a pure set X is defined to be the least upper bound of all successors of ranks of members of X. For example, the empty set is assigned rank 0, while the set {{}} containing only the empty set is assigned rank 1. For each ordinal á, the set Vá is defined to consist of all pure sets with rank less than á. The entire von Neumann universe is denoted V.

In set theory and related branches of mathematics, the von Neumann universe, or von Neumann hierarchy of sets, denoted V, is the class of hereditary well-founded sets. This collection, which is formalized by Zermelo–Fraenkel set theory (ZFC), is often used to provide an interpretation or motivation of the axioms of ZFC. The rank of a well-founded set is defined inductively as the smallest ordinal number greater than the ranks of all members of the set. In particular, the rank of the empty set is zero, and every ordinal has a rank equal to itself. The sets in V are divided into the transfinite hierarchy Vá, called the cumulative hierarchy, based on their rank.

Applications and interpretations

Applications of V as Models for set Theories

If ù is the set of natural numbers, then Vw is the set of hereditarily finite sets, which is a model of set theory without the axiom of infinity. V_{w+w} is the universe of "ordinary mathematics", and is a model of Zermelo set theory. A simple argument in favour of the adequacy of V_{w+w} is the observation that V_{w+1} is adequate for the integers, while V_{w+2} is adequate for the real numbers, and most other normal mathematics can be built as relations of various kinds from these sets without needing the axiom of replacement to go outsideV_{w+w}. If ê is an inaccessible cardinal, then VK is a model of Zermelo–Fraenkel set theory (ZFC) itself, and VK+1 is a model of Morse–Kelley set theory. (Note that every ZFC model is also a ZF model, and every ZF model is also a Z model.)

Interpretation of V as the "set of all sets"

V is not "the set of all sets" for two reasons. First, it is not a set; although each individual stage Vá is a set, their union V is a proper class. Second, the sets in V are only the well-founded sets. The axiom of foundation (or regularity) demands that every set be well founded and hence in V, and thus in ZFC every set is in V. But other axiom systems may omit the axiom of foundation or replace it by a strong negation (an example is Aczel's anti-foundation axiom). These non-well-founded set theories are not commonly employed, but are still possible to study. A third objection to the "set of all sets" interpretation is that not all sets are necessarily "pure sets", which are constructed from the empty set using power sets and unions. Zermelo proposed in 1908 the inclusion of urelements, from which he constructed a transfinite recursive hierarchy in 1930. Such urelements are used extensively in model theory, particularly in Fraenkel-Mostowski models.

V and the Axiom of Regularity

The formula V = Ã”áVá is often considered to be a theorem, not a definition. Roitman states (without references) that the realization that the axiom of regularity is equivalent to the equality of the universe of ZF sets to the cumulative hierarchy is due to von Neumann.

The Existential Status of V

Since the class V may be considered to be the arena for most of mathematics, it is important to establish that it "exists" in some sense. Since existence is a difficult concept, one typically replaces the existence question with the consistency question, that is, whether the concept is free of contradictions. A major obstacle is posed by Gödel's incompleteness theorems, which effectively imply the impossibility of proving the consistency of ZF set theory in ZF set theory itself, provided that it is in fact consistent.

The integrity of the von Neumann universe depends fundamentally on the integrity of the ordinal numbers, which act as the rank parameter in the construction, and the integrity of transfinite induction, by which both the ordinal numbers and the von Neumann universe are constructed. The integrity of the ordinal number construction may be said to rest upon von Neumann's 1923 and 1928 papers. The integrity of the construction of V by transfinite induction may be said to have then been established in Zermelo's 1930 paper.

History

The cumulative type hierarchy, also known as the von Neumann universe, is claimed by Gregory H. Moore (1982) to be inaccurately attributed to von Neumann.

The first publication of the von Neumann universe was by Ernst Zermelo in 1930. Existence and uniqueness of the general transfinite recursive definition of sets was demonstrated in 1928 by von Neumann for both Zermelo-Fraenkel set theory and Neumann's own set theory (which later developed into NBG set theory). In neither of these papers did he apply his transfinite recursive method to construct the universe of all sets. The presentations of the von Neumann universe by Bernays and Mendelson both give credit to von Neumann for the transfinite induction construction method, although not for its application to the construction of the universe of ordinary sets.

The notation V is not a tribute to the name of von Neumann. It was used for the universe of sets in 1889 by Peano, the letter V signifying "Verum", which he used both as a logical symbol and to denote the class of all individuals. Peano's notation V was adopted also by Whitehead and Russell for the class of all sets in 1910. The V notation (for the class of all sets) was not used by von Neumann in his 1920s papers about ordinal numbers and transfinite induction. Paul Cohen explicitly attributes his use of the letter V (for the class of all sets) to a 1940 paper by Gödel, although Gödel most likely obtained the notation from earlier sources such as Whitehead and Russell.

Philosophical Perspectives

There are two approaches to understanding the relationship of the von Neumann universe V to ZFC (along with many variations of each approach, and shadings between them). Roughly, formalists will tend to view V as something that flows from the ZFC axioms (for example, ZFC proves that every set is in V). On the other hand, realists are more likely to see the von Neumann hierarchy as something directly accessible to the intuition, and the axioms of ZFC as propositions for whose truth in V we can give direct intuitive arguments in natural language. A possible middle position is that the mental picture of the von Neumann hierarchy provides the ZFC axioms with a motivation (so that they are not arbitrary), but does not necessarily describe objects with real existence.

AXIOMATIC SET THEORY

Elementary set theory can be studied informally and intuitively, and so can be taught in primary schools using Venn diagrams. The intuitive approach tacitly assumes that a set may be formed from the class of all objects satisfying any particular defining condition. This assumption gives rise to paradoxes, the simplest and best known of which are Russell's paradox and the Burali-Forti paradox. Axiomatic set theory was originally devised to rid set theory of such paradoxes.

The most widely studied systems of axiomatic set theory imply that all sets form a cumulative hierarchy. Such systems come in two flavors, those whose ontology consists of:

- Sets alone. This includes the most common axiomatic set theory, Zermelo–Fraenkel set theory (ZFC), which includes the axiom of choice. Fragments of ZFC include:
- Zermelo set theory, which replaces the axiom schema of replacement with that of separation;
- General set theory, a small fragment of Zermelo set theory sufficient for the Peano axioms and finite sets;
- Kripke–Platek set theory, which omits the axioms of infinity, powerset, and choice, and weakens the axiom schemata of separation and replacement.
- Sets and proper classes. These include Von Neumann–Bernays–Gödel set theory, which has the same strength as ZFC for theorems about sets alone, and Morse–Kelley set theory and Tarski–Grothendieck set theory, both of which are stronger than ZFC.

The above systems can be modified to allow urelements, objects that can be members of sets but that are not themselves sets and do not have any members. The systems of New Foundations NFU (allowing urelements) and NF (lacking them) are not based on a cumulative hierarchy. NF and NFU include a "set of everything, " relative to which produces sets for which the axiom of choice does not hold. Systems of constructive set theory, such as CST, CZF, and IZF, embed their set axioms in intuitionistic instead of classical logic. Yet other systems accept classical logic but feature a nonstandard membership relation. These include rough set theory and fuzzy set theory, in which the value of an atomic formula embodying the membership relation is not simply True or False. The Boolean-valued models of ZFC are a related subject.An enrichment of ZFC called internal set theory was proposed by Edward Nelson in 1977.

Applications

and propositional Many mathematical concepts can be defined precisely using only set theoretic concepts. For example, mathematical structures as diverse as graphs, manifolds, rings, and vector spaces can all be defined as sets satisfying various (axiomatic) properties. Equivalence and order relationsare ubiquitous in mathematics, and the theory of mathematical relations can be described in set theory. Set theory is also a promising foundational system for much of mathematics. Since the publication of the first volume of Principia Mathematica, it has been claimed that most or even all mathematical theorems can be derived using an aptly designed set of axioms for set theory, augmented with many definitions, using first or second order logic. For example, properties of the natural and real numbers can be derived within set theory, as each number system can be identified with a set of equivalence classes under a suitable equivalence relation whose field is some infinite set.

Set theory as a foundation for mathematical analysis, topology, abstract algebra, and discrete mathematics is likewise uncontroversial; mathematicians accept that (in principle) theorems in these areas can be derived from the relevant definitions and the axioms of set theory. Few full derivations of complex mathematical theorems from set theory have been formally verified, however, because such formal derivations are often much longer than the natural language proofs mathematicians commonly present. One verification project, Metamath, includes human-written, computer-verified derivations of more than 12,000 theorems starting from ZFC set theory, first order logiclogic

Areas of Study

Set theory is a major area of research in mathematics, with many interrelated subfields.

Combinatorial set Theory

Combinatorial set theory concerns extensions of finite combinatorics to infinite sets. This includes the study of cardinal arithmetic and the study of extensions of Ramsey's theorem such as the Erdõs–Rado theorem.

Combinatorics

Combinatorics is an area of mathematics primarily concerned with counting, both as a means and an end in obtaining results, and certain properties of finite structures. It is closely related to many other areas of mathematics and has many applications ranging from logic to statistical physics, from evolutionary biology to computer science, etc.To fully understand the scope of combinatorics requires a great deal of further amplification, the details of which are not universally agreed upon. According to H.J. Ryser, a definition of the subject is difficult because it crosses so many mathematical subdivisions. Insofar as an area can be described by the types of problems it addresses, combinatorics is involved with the enumeration (counting) of specified structures, sometimes referred to as arrangements

- Or configurations in a very general sense, associated with finite systems,
- The existence of such structures that satisfy certain given criteria,

- The construction of these structures, perhaps in many ways, and
- Optimization, finding the "best" structure or solution among several possibilities, be it the "largest", "smallest" or satisfying some other optimality criterion.

Leon Mirsky has said: "combinatorics is a range of linked studies which have something in common and yet diverge widely in their objectives, their methods, and the degree of coherence they have attained." One way to define combinatorics is, perhaps, to describe its subdivisions with their problems and techniques. This is the approach that is used below. However, there are also purely historical reasons for including or not including some topics under the combinatorics umbrella. Although primarily concerned with finite systems, some combinatorial questions and techniques can be extended to an infinite (specifically, countable) but discrete setting. Combinatorics is well known for the breadth of the problems it tackles. Combinatorial problems arise in many areas of pure mathematics, notably in algebra, probability theory, topology, and geometry, as well as in its many application areas. Many combinatorial questions have historically been considered in isolation, giving an ad hoc solution to a problem arising in some mathematical context. In the later twentieth century, however, powerful and general theoretical methods were developed, making combinatorics into an independent branch of mathematics in its own right. One of the oldest and most accessible parts of combinatorics is graph theory, which by itself has numerous natural connections to other areas. Combinatorics is used frequently in computer science to obtain formulas and estimates in the analysis of algorithms. A mathematician who studies combinatorics is called a combinatorialist.

History

An example of change ringing (with six bells), one of the earliest nontrivial results in graph theory.Basic combinatorial concepts and enumerative results appeared throughout the ancient world. In the 6th century BCE, ancient Indian physician Sushruta asserts in Sushruta Samhita that 63 combinations can be made out of 6 different tastes, taken one at a time, two at a time, etc., thus computing all 26 " 1 possibilities. Greek historian Plutarch discusses an argument between Chrysippus (3rd century BCE) and Hipparchus (2nd century BCE) of a rather delicate enumerative problem, which was later shown to be related to Schröder–Hipparchus numbers. In the Ostomachion, Archimedes (3rd century BCE) considers a tiling puzzle. In the Middle Ages, combinatorics continued to be studied, largely outside of the European civilization. The Indian mathematician Mahâvîra (c. 850) provided formulae for the number of permutations and combinations, and these formulas may have been familiar to Indian mathematicians as early as the 6th century CE. The philosopher and astronomer Rabbi Abraham ibn Ezra (c. 1140) established the symmetry of binomial coefficients, while a closed formula was obtained later by the talmudist and mathematician Levi ben Gerson (better known as Gersonides), in 1321. The arithmetical triangle— a graphical diagram showing relationships among the binomial coefficients— was presented by mathematicians in treatises dating as far back as the 10th century, and would eventually become known as Pascal's triangle. Later, in Medieval England, campanology provided examples of what is now known as Hamiltonian cycles in certain Cayley graphs on permutations.

During the Renaissance, together with the rest of mathematics and the sciences, combinatorics enjoyed a rebirth. Works of Pascal, Newton, Jacob Bernoulli and Euler became foundational in the emerging field. In modern times, the works of J.J. Sylvester (late 19th century) and Percy MacMahon (early 20th century) helped lay the foundation for enumerative and algebraic combinatorics. Graph theory also enjoyed an explosion of interest at the same time, especially in connection with the four color problem. In the second half of the 20th century, combinatorics enjoyed a rapid growth, which led to establishment of dozens of new journals and conferences in the subject. In part, the growth was spurred by new connections and applications to other fields, ranging from algebra to probability, from functional analysis to number theory, etc. These connections shed the boundaries between combinatorics and parts of mathematics and theoretical computer science, but at the same time led to a partial fragmentation of the field.

APPROACHES AND SUBFIELDS OF COMBINATORICS

Enumerative Combinatorics

Five binary trees on three vertices, an example of Catalan numbers.Enumerative combinatorics is the most classical area of combinatorics and concentrates on counting the number of certain combinatorial objects. Although counting the number of elements in a set is a rather broad mathematical problem, many of the problems that arise in applications have a relatively simple combinatorial description. Fibonacci numbers is the basic example of a problem in enumerative combinatorics. The twelvefold way provides a unified framework for counting permutations, combinations and partitions.

Analytic Combinatorics

Analytic combinatorics concerns the enumeration of combinatorial structures using tools from complex analysis and probability theory. In contrast with enumerative combinatorics, which uses explicit combinatorial formulae and generating functions to describe the results, analytic combinatorics aims at obtaining asymptotic formulae.

Partition Theory

A plane Partition.

Partition theory studies various enumeration and asymptotic problems related to integer partitions, and is closely related to q-series, special functions and orthogonal polynomials. Originally a part of number theory and analysis, it is now considered a part of combinatorics or an independent field. It incorporates the bijective approach and various tools in analysis and analytic number theory, and has connections with statistical mechanics.

GRAPH THEORY

Petersen Graph.

Graphs are basic objects in combinatorics. The questions range from counting (e.g., the number of graphs on n vertices with k edges) to structural (e.g., which graphs contain Hamiltonian cycles) to algebraic questions (e.g., given a graph G and two numbers x and y, does the Tutte polynomial TG(x,y) have a combinatorial interpretation?). Although there are very strong connections between graph theory and combinatorics, these two are sometimes thought of as separate subjects. This is due to the fact that while combinatorial methods apply to many graph theory problems, the two are generally used to seek solutions to different problems.

Design Theory

Design theory is a study of combinatorial designs, which are collections of subsets with certain intersection properties. Block designs are combinatorial designs of a special type. This area is one of the oldest parts of combinatorics, such as in Kirkman's schoolgirl problem proposed in 1850. The solution of the problem is a special case of a Steiner system, which systems play an important role in the classification of finite simple groups. The area has further connections to coding theory and geometric combinatorics.

Finite Geometry

Finite geometry is the study of geometric systems having only a finite number of points. Structures analogous to those found in continuous geometries (Euclidean plane, real projective space, etc.) but defined combinatorially are the main items studied. This area provides a rich source of examples for design theory. It should not be confused with discrete geometry (combinatorial geometry).

Order Theory

Hasse diagram of the powerset of {x,y,z} ordered by inclusion. Order theory is the study of partially ordered sets, both finite and infinite. Various examples of partial orders appear in algebra, geometry, number theory and throughout combinatorics and graph theory. Notable classes and examples of partial orders include lattices and Boolean algebras.

Matroid Theory

Matroid theory abstracts part of geometry. It studies the properties of sets (usually, finite sets) of vectors in a vector space that do not depend on the particular coefficients in a linear dependence relation. Not only the structure but also enumerative properties belong to matroid theory. Matroid theory was introduced by Hassler Whitney and studied as a part of order theory. It is now an independent field of study with a number of connections with other parts of combinatorics.

Extremal Combinatorics

Extremal combinatorics studies extremal questions on set systems. The types of questions addressed in this case are about the largest possible graph which satisfies certain properties. For example, the largest triangle-free graph on 2n vertices is a complete bipartite graph Kn,n. Often it is too hard even to find the extremal answer f(n) exactly and one can only give an asymptotic estimate. Ramsey theory is another part of extremal combinatorics. It states that any sufficiently large configuration will contain some sort of order. It is an advanced generalization of the pigeonhole principle.

Probabilistic combinatorics

Self-Avoiding Walk in a Square grid Graph.

In probabilistic combinatorics, the questions are of the following type: what is the probability of a certain property for a random discrete object, such as a random graph? For instance, what is the average number of triangles in a random graph? Probabilistic methods are also used to determine the existence of combinatorial objects with certain prescribed properties (for which explicit examples might be difficult to find), simply by observing that the probability of randomly selecting an object with those properties is greater than 0. This approach (often referred to as the probabilistic method) proved highly effective in applications to extremal combinatorics and graph theory. A closely related area is the study of finite Markov chains, especially on combinatorial objects. Here again probabilistic tools are used to estimate the mixing time. Often associated with Paul Erdõs, who did the pioneering work on the subject, probabilistic combinatorics was traditionally viewed as a set of tools to study problems in other parts of combinatorics. However, with the growth of applications to analysis of algorithms in computer science computer science, as well as classical probability, additive and probabilistic number theory, the area recently grew to become an independent field of combinatorics.

Algebraic Combinatorics

- Young Diagram of a Partition (5,4,1).

Algebraic combinatorics is an area of mathematics that employs methods of abstract algebra, notably group theory and representation theory, in various combinatorial contexts and, conversely, applies combinatorial techniques to problems in algebra. Algebraic combinatorics is continuously expanding its scope, in both topics and techniques, and can be seen as the area of mathematics where the interaction of combinatorial and algebraic methods is particularly strong and significant.

Combinatorics on Words

Construction of a Thue–Morse Infinite Word.

Combinatorics on words deals with formal languages. It arose independently within several branches of mathematics, including number theory, group theory and probability. It has applications to enumerative combinatorics, fractal analysis, theoretical computer science, automata theory and linguistics. While many applications are new, the classical Chomsky–Schützenberger hierarchy of classes of formal grammars is perhaps the best-known result in the field.

GEOMETRIC COMBINATORICS

An Icosahedron

Geometric combinatorics is related to convex and discrete geometry, in particular polyhedral combinatorics. It asks, for example, how many faces of each dimension a convex polytope can have. Metric properties of polytopes play an important role as well, e.g. the Cauchy theorem on the rigidity of convex polytopes. Special polytopes are also considered, such as permutohedra, associahedra and Birkhoff polytopes. We should note that combinatorial geometry is an old fashioned name for discrete geometry.

Topological Combinatorics

Splitting a Necklace with Two cuts.

Combinatorial analogs of concepts and methods in topology are used to study graph coloring, fair division, partitions, partially ordered sets, decision trees, necklace problems and discrete Morse theory. It should not be confused with combinatorial topology which is an older name for algebraic topology.

Arithmetic Combinatorics

Arithmetic combinatorics arose out of the interplay between number theory, combinatorics, ergodic theory and harmonic analysis. It is about combinatorial estimates associated with arithmetic operations (addition, subtraction, multiplication, and division). Additive number theory (sometimes also called additive combinatorics) refers to the special case when only the operations of addition and subtraction are involved. One important technique in arithmetic combinatorics is the ergodic theory of dynamical systems.

Infinitary Combinatorics

Infinitary combinatorics, or combinatorial set theory, is an extension of ideas in combinatorics to infinite sets. It is a part of set theory, an area of mathematical logic, but uses tools and ideas from both set theory and extremal combinatorics. Gian-Carlo Rota used the name continuous combinatorics to describe geometric probability, since there are many analogies between counting and measure.

Related fields

Kissing spheres are connected to both coding theory and discrete geometry.

Combinatorial Optimization

Combinatorial optimization is the study of optimization on discrete and combinatorial objects. It started as a part of combinatorics and graph theory, but is now viewed as a branch of applied mathematics and computer science, related to operations research, algorithm theory and computational complexity theory.

Coding Theory

Coding theory started as a part of design theory with early combinatorial constructions of error-correcting codes. The main idea of the subject is to design efficient and reliable methods of data transmission. It is now a large field of study, part of information theory.

Discrete and Computational Geometry

Discrete geometry (also called combinatorial geometry) also began as a part of combinatorics, with early results on convex polytopes and kissing numbers. With the emergence of applications of discrete geometry to computational geometry, these two fields partially merged and became a separate field of study. There remain many connections with geometric and topological combinatorics, which themselves can be viewed as outgrowths of the early discrete geometry.

Combinatorics and Dynamical Systems

Combinatorial aspects of dynamical systems is another emerging field. Here dynamical systems can be defined on combinatorial objects. See for example graph dynamical system.

Combinatorics and Physics

There are increasing interactions between combinatorics and physics, particularly statistical physics. Examples include an exact solution of the Ising model, and a connection between the Potts model on one hand, and the chromatic and Tutte polynomials on the other hand.

Extremal Combinatorics

Extremal combinatorics is a field of combinatorics, which is itself a part of mathematics. Extremal combinatorics studies how large or how small a collection of finite objects (numbers, graphs, vectors, sets, etc.) can be, if it has to satisfy certain restrictions. Much of extremal combinatorics concerns classes of sets; this is called extremal set theory. For instance, in an n-element set, what is the largest number of k-element subsets that can pairwise intersect one another? What is the largest number of subsets of which none contains any other? The latter question is answered by Sperner's theorem, which gave rise to much of extremal set theory.

Another kind of example: How many people can we invite to a party where among each three people there are two who know each other and two who don't know each other? Ramsey theory shows that at most five persons can attend such a party. Or, suppose we are given a finite set of nonzero integers, and are asked to mark as large a subset as possible of this set under the restriction that the sum of any two marked integers cannot be marked. It appears that (independent of what the given integers actually are!) we can always mark at least one-third of them.

Descriptive set Theory

Descriptive set theory is the study of subsets of the real line and, more generally, subsets of Polish spaces. It begins with the study of pointclasses in the Borel hierarchy and extends to the study of more complex hierarchies such as the projective hierarchy and the Wadge hierarchy. Many properties of Borel sets can be established in ZFC, but proving these properties hold for more complicated sets requires additional axioms related to determinacy and large cardinals. The field of effective descriptive set theory is between set theory and recursion theory. It includes the study of lightface pointclasses, and is closely related to hyperarithmetical theory. In many cases, results of classical descriptive set theory have effective versions; in some cases, new results are obtained by proving the effective version first and then extending ("relativizing") it to make it more broadly applicable. A recent area of research concerns Borel equivalence relations and more complicated definable equivalence relations. This has important applications to the study of invariants in many fields of mathematics.

Descriptive set Theory

In mathematical logic, descriptive set theory (DST) is the study of certain classes of "well-behaved" subsets of the real line and other Polish spaces. As well as being one of the primary areas of research in set theory, it has applications to other areas of mathematics such as functional analysis, ergodic theory, the study of operator algebras and group actions, and mathematical logic.

Polish Spaces

Descriptive set theory begins with the study of Polish spaces and their Borel sets. A Polish space is a second-countable topological space that is metrizable with a complete metric. Equivalently, it is a complete separable metric space whose metric has been "forgotten". Examples include the real line the Baire space the Cantor space and the Hilbert cube I N

Universality Properties

The class of Polish spaces has several universality properties, which show that there is no loss of generality in considering Polish spaces of certain restricted forms.

- Every Polish space is homeomorphic to a Gä subspace of the Hilbert cube, and every Gä subspace of the Hilbert cube is Polish.
- Every Polish space is obtained as a continuous image of Baire space; in fact every Polish space is the image of a continuous bijection defined on a closed subset of Baire space. Similarly, every compact Polish space is a continuous image of Cantor space.

Because of these universality properties, and because the Baire space has the convenient property that it is homeomorphic to many results in descriptive set theory are proved in the context of Baire space alone.

Borel Sets

The class of Borel sets of a topological space X consists of all sets in the smallest ó-algebra containing the open sets of X. This means that the Borel sets of X are the smallest collection of sets such that:

- Every open subset of X is a Borel set.
- If A is a Borel set, so is X " That is, the class of Borel sets are closed under complementation.
- If An is a Borel set for each natural number n, then the union Ã" A is a Borel set. That is, the Borel sets are closed under countable unions.

A fundamental result shows that any two uncountable Polish spaces X and Y are Borel isomorphic: there is a bijection from X to Y such that the preimage of any Borel set is Borel, and the image of any Borel set is Borel. This gives additional justification to the practice of restricting attention to Baire space and Cantor space, since these and any other Polish spaces are all isomorphic at the level of Borel sets.

Borel Hierarchy

Each Borel set of a Polish space is classified in the Borel hierarchy based on how many times the operations of countable union and complementation must be used to obtain the set, beginning from open sets. The classification is in terms of countable ordinal numbers. For each nonzero countable ordinal á there are classes.

- Every open set is declared to be.
- A set is declared to be if and only if its complement is.
- A set A is declared to be, $\delta > 1$, if there is a sequence $\{A_i\}$ of sets, each of which is for some $\lambda(i) < \delta$, such that.
- A set is Ä á if and only if it is both A theorem shows that any set that is or is, and any set is both and for all. Thus the hierarchy has the following structure, where arrows indicate inclusion.

Regularity Properties of Borel Sets

Classical descriptive set theory includes the study of regularity properties of Borel sets. For example, all Borel sets of a Polish space have the property of Baire and the perfect set property. Modern descriptive set theory includes the study of the ways in which these results generalize, or fail to generalize, to other classes of subsets of Polish spaces.

Analytic and Coanalytic sets

Just beyond the Borel sets in complexity are the analytic sets and coanalytic sets. A subset of a Polish space X is analytic if it is the continuous image of a Borel subset of some other Polish space. Although any continuous preimage of a Borel set is Borel, not all analytic sets are Borel sets. A set is coanalytic if its complement is analytic.

Projective Sets and Wadge Degrees

Many questions in descriptive set theory ultimately depend upon set-theoretic considerations and the properties of ordinal and cardinal numbers. This phenomenon is particularly apparent in the projective sets. These are defined via the projective hierarchy on a Polish space X:

- A set is declared to be if it is analytic.
- A set is if it is coanalytic.
- A set A is if there is a subset B of such that A is the projection of B to the first coordinate.
- A set A is if there is a subset B of such that A is the projection of B to the first coordinate.
- A set is if it is both and.

The properties of the projective sets are not completely determined by ZFC. Under the assumption V = L, not all projective sets have the perfect set property or the property of Baire. However,under the assumption of projective determinacy, all projective sets have both the perfect set property and the property of Baire. This is related to the fact that ZFC proves Borel determinacy, but not projective determinacy. More generally, the entire collection of sets of elements of a Polish space X can be grouped into equivalence classes, known as Wadge degrees, that generalize the projective hierarchy. These degrees are ordered in the Wadge hierarchy. The axiom of determinacy implies that the Wadge hierarchy on any Polish space is well-founded and of length È, with structure extending the projective hierarchy.

Borel Equivalence Relations

A contemporary area of research in descriptive set theory studies Borel equivalence relations. A Borel equivalence relation on a Polish space X is a Borel subset of that is an equivalence relation on X.

Effective Descriptive set Theory

The area of effective descriptive set theory combines the methods of descriptive set theory with those of generalized recursion theory (especially hyperarithmetical theory). In particular, it focuses on lightface analogues of hierarchies of classical descriptive set theory. Thus the hyperarithmetic hierarchy is studied instead of the Borel hierarchy, and the analytical hierarchy instead of the projective hierarchy. This research is related to weaker versions of set theory such as Kripke–Platek set theory and second-order arithmetic.

Fuzzy set Theory

In set theory as Cantor defined and Zermelo and Fraenkel axiomatized, an object is either a member of a set or not. In fuzzy set theory this condition was relaxed by Lotfi A. Zadeh so an object has a degree of membership in a set, a number between 0 and 1. For example, the degree of membership of a person in the set of "tall people" is more flexible than a simple yes or no answer and can be a real number such as 0.75.

Most of our traditional tools for formal modeling,reasoning, and computing are crisp, determin-istic, and precise in character. Crisp means dichoto-mous, that is, yes-or-no type rather than more-or-lesstype. In traditional dual logic, for instance, a state-ment can be true or false—and nothing in between.In set theory, an element can either belong to a setor not; in optimization a solution can be feasibleor not. Precision assumes that the parameters of amodel represent exactly the real system that has beenmodeled. This, generally, also implies that the modelis unequivocal, that is, that it contains no ambigui-ties. Certainty eventually indicates that we assume thestructures and parameters of the model to be definitelyknown and that there are no doubts about their val-ues or their occurrence. Unluckily these assumptionsand beliefs are not justified if it is important, that themodel describes well reality (which is neither crispnor certain). In addition, the complete description ofa real system would often require far more detaileddata than a human being could ever recognize simul-taneously, process, and understand. This situation hasalready been recognized by thinkers in the past. In1923, the philosopher B. Russell referred to the firstpoint when he wrote: 'All traditional logic habitually assumes that precise symbols are being employed. Itis, therefore, not applicable to this terrestrial life butonly to an imagined celestial existence'.1L. Zadehreferred to the second point, when he wrote: 'As thecomplexity of a system increases, our ability to makeprecise and yet significant statements about its behav-ior diminishes until a threshold is reached beyondwhich precision and significance (or relevance) becomealmost mutually exclusive characteristics'.2For a longtime, probability theory and statistics have been thepredominant theories and tools to model uncertaintiesof reality.3,4They are based—as all formal theo-ries—on certain axiomatic assumptions, which arehardly ever tested, when these theories are applied toreality. In the meantime more than 20 other 'uncer-tainty theories' have been developed,5which partlycontradict each other and partly complement eachother. Fuzzy set theory—formally speaking—is oneof these theories, which was initially intended to bean extension of dual logic and/or classical set theory. During the last decades, it has been developed in thedirection of a powerful 'fuzzy' mathematics. Whenit is used, however, as a tool to model reality betterthan traditional theories, then an empirical validationis very desirable.6In the following sections, the for-mal theory is described and some of the attemptsto verify the theory empirically in reality will besummarized.

HISTORY OF FUZZY SET THEORY

A short historical review may be useful to betterunderstand the character and motivation of thistheory. The first publications in fuzzy set theoryby Zadeh7and Goguen8show the intention of theauthors to generalize the classical notion of a setand a proposition to accommodate fuzziness in thesense that it is contained in human language, that is,in human judgment, evaluation, and decisions. Zadehwrites: 'The notion of a fuzzy set provides a convenientpoint of departure for the construction of a conceptualframework which parallels in many respects theframework used in the case of ordinary sets, butis more general than the latter and, potentially, mayprove to have a much wider scope of applicability,particularly in the fields of pattern classification andinformation processing. Essentially, such a frameworkprovides a natural way of dealing with problems inwhich the source of imprecision is the absence ofsharply defined criteria of class membership ratherthan the presence of random variables'.7'Imprecision'here is meant in the sense of vagueness rather thanthe lack of knowledge about the value of a parameter(as in tolerance analysis). Fuzzy set theory provides astrict mathematical framework (there is nothing fuzzyabout fuzzy set theory!) in which vague conceptualphenomena can be precisely and rigorously studied.It can also be considered as a modeling language,well suited for situations in which fuzzy relations,criteria, and phenomena exist. The acceptance ofthis theory grew slowly in the 1960s and 1970s ofthe last century. In the second half of the 1970s,however, the first successful practical applications inthe control of technological processes via fuzzy rule-based systems, called fuzzy control (heating systems,cement factories, etc.), boosted the interest in this areaconsiderably. Successful applications, particularly inJapan, in washing machines, video cameras, cranes,subway trains, and so on triggered further interestand research in the 1980s so that in 1984 alreadyapproximately 4000 publications existed and in 2000more than 30,000. Roughly speaking, fuzzy set theoryduring the last decades has developed along twolines

1.As a formal theory that, when maturing, becamemore sophisticated and specified and wasenlarged by original ideas and concepts as wellas by 'embracing' classical mathematical areas,such as algebra,9,10graph theory,11topology,and so on by generalizing or 'fuzzifying' them.This development is still ongoing.2.As an application-oriented 'fuzzy technology',that is, as a tool for modeling, problem solving,and data mining that has been proven superior toexisting methods in many cases and as attractive'add-on' to classical approaches in othercases.In 1992, in three simultaneous conferencesin Europe, Japan, and the United States, thethree areas of fuzzy set theory, neural nets, andevolutionary computing (genetic algorithms) joinedforces and are henceforth known under 'compu-tational intelligence'.8,12,13In a similar way, theterm 'soft computing' is used for a number ofapproaches that deal essentially with uncertainty andimprecision.

Inner Model Theory

An inner model of Zermelo–Fraenkel set theory (ZF) is a transitive class that includes all the ordinals and satisfies all the axioms of ZF. The canonical example is the constructible universeL developed by Gödel. One reason that the study of inner models is of interest is that it can be used to prove consistency results. For example, it can be shown that regardless of whether a model V of ZF satisfies the continuum hypothesis or the axiom of choice, the inner model L constructed inside the original model will satisfy both the generalized continuum hypothesis and the axiom of choice. Thus the assumption that ZF is consistent (has at least one model) implies that ZF together with these two principles is consistent.

The study of inner models is common in the study of determinacy and large cardinals, especially when considering axioms such as the axiom of determinacy that contradict the axiom of choice. Even if a fixed model of set theory satisfies the axiom of choice, it is possible for an inner model to fail to satisfy the axiom of choice. For example, the existence of sufficiently large cardinals implies that there is an inner model satisfying the axiom of determinacy (and thus not satisfying the axiom of choice).

Large Cardinals

A large cardinal is a cardinal number with an extra property. Many such properties are studied, including inaccessible cardinals, measurable cardinals, and many more. These properties typically imply the cardinal number must be very large, with the existence of a cardinal with the specified property unprovable in Zermelo-Fraenkel set theory.

Large Cardinal

In the mathematical field of set theory, a large cardinal property is a certain kind of property of transfinite cardinal numbers. Cardinals with such properties are, as the name suggests, generally very "large" (for example, bigger than the least á such that á=ùá). The proposition that such cardinals exist cannot be proved in the most common axiomatization of set theory, namely ZFC, and such propositions can be viewed as ways of measuring how "much", beyond ZFC, one needs to assume to be able to prove certain desired results. In other words, they can be seen, in Dana Scott's phrase, as quantifying the fact "that if you want more you have to assume more". There is a rough convention that results provable from ZFC alone may be stated without hypotheses, but that if the proof requires other assumptions (such as the existence of large cardinals), these should be stated. Whether this is simply a linguistic convention, or something more, is a controversial point among distinct philosophical schools (see Motivations and epistemic status below). A large cardinal axiom is an axiom stating that there exists a cardinal (or perhaps many of them) with some specified large cardinal property.

Most working set theorists believe that the large cardinal axioms that are currently being considered are consistent with ZFC. These axioms are strong enough to imply the consistency of ZFC. This has the consequence (via Gödel's second incompleteness theorem) that their consistency with ZFC cannot be proven in ZFC (assuming ZFC is consistent).

There is no generally agreed precise definition of what a large cardinal property is, though essentially everyone agrees that those in the list of large cardinal properties are large cardinal properties.

Partial Definition

A necessary condition for a property of cardinal numbers to be a large cardinal property is that the existence of such a cardinal is not known to be inconsistent with ZFC and it has been proven that if ZFC is consistent, then ZFC + "no such cardinal exists" is consistent.

Hierarchy of Consistency Strength

A remarkable observation about large cardinal axioms is that they appear to occur in strict linear order by consistency strength. That is, no exception is known to the following: Given two large cardinal axioms A1 and A2, exactly one of three things happens:

1. ZFC proves "ZFC+A1 is consistent if and only if ZFC+A2 is consistent";
2. ZFC+A1 proves that ZFC+A2 is consistent; or
3. ZFC+A2 proves that ZFC+A1 is consistent.

These are mutually exclusive, unless one of the theories in question is actually inconsistent. In case 1 we say that A1 and A2 are equiconsistent. In case 2, we say that A1 is consistency-wise stronger than A2 (vice versa for case 3). If A2 is stronger than A1, then ZFC+A1 cannot prove ZFC+A2 is consistent, even with the additional hypothesis that ZFC+A1 is itself consistent (provided of course that it really is). This follows from Gödel's second incompleteness theorem.

The observation that large cardinal axioms are linearly ordered by consistency strength is just that, an observation, not a theorem. (Without an accepted definition of large cardinal property, it is not subject to proof in the ordinary sense). Also, it is not known in every case which of the three cases holds. Saharon Shelah has asked, "[i]s there some theorem explaining this, or is our vision just more uniform than we realize?" Woodin, however, deduces this from the Ù-conjecture, the main unsolved problem of his Ù-logic. It is also noteworthy that many combinatorial statements are exactly equiconsistent with some large cardinal rather than, say, being intermediate between them.

It should also be noted that the order of consistency strength is not necessarily the same as the order of the size of the smallest witness to a large cardinal axiom. For example, the existence of a huge cardinal is much stronger, in terms of consistency strength, than the existence of a supercompact cardinal, but assuming both exist, the first huge is smaller than the first supercompact.

Motivations and Epistemic Status

Large cardinals are understood in the context of the von Neumann universe V, which is built up by transfinitely iterating the powerset operation, which collects together all subsets of a given

set. Typically, models in which large cardinal axioms fail can be seen in some natural way as submodels of those in which the axioms hold. For example, if there is an inaccessible cardinal, then "cutting the universe off" at the height of the first such cardinal yields a universe in which there is no inaccessible cardinal. Or if there is a measurable cardinal, then iterating the definable powerset operation rather than the full one yields Gödel's constructible universe, L, which does not satisfy the statement "there is a measurable cardinal" (even though it contains the measurable cardinal as an ordinal). Thus, from a certain point of view held by many set theorists (especially those inspired by the tradition of the Cabal), large cardinal axioms "say" that we are considering all the sets we're "supposed" to be considering, whereas their negations are "restrictive" and say that we're considering only some of those sets. Moreover the consequences of large cardinal axioms seem to fall into natural patterns (see Maddy, "Believing the Axioms, II"). For these reasons, such set theorists tend to consider large cardinal axioms to have a preferred status among extensions of ZFC, one not shared by axioms of less clear motivation (such as Martin's axiom) or others that they consider intuitively unlikely (such as V = L). The hardcore realists in this group would state, more simply, that large cardinal axioms are true.

This point of view is by no means universal among set theorists. Some formalists would assert that standard set theory is by definition the study of the consequences of ZFC, and while they might not be opposed in principle to studying the consequences of other systems, they see no reason to single out large cardinals as preferred. There are also realists who deny that ontological maximalism is a proper motivation, and even believe that large cardinal axioms are false. And finally, there are some who deny that the negations of large cardinal axioms are restrictive, pointing out that (for example) there can be a transitive set model in L that believes there exists a measurable cardinal, even though L itself does not satisfy that proposition.

Determinacy

Determinacy refers to the fact that, under appropriate assumptions, certain two-player games of perfect information are determined from the start in the sense that one player must have a winning strategy. The existence of these strategies has important consequences in descriptive set theory, as the assumption that a broader class of games is determined often implies that a broader class of sets will have a topological property. The axiom of determinacy (AD) is an important object of study; although incompatible with the axiom of choice, AD implies that all subsets of the real line are well behaved (in particular, measurable and with the perfect set property). AD can be used to prove that the Wadge degrees have an elegant structure.

Determinacy

Determinacy is a subfield of set theory, a branch of mathematics, that examines the conditions under which one or the other player of a game has a winning strategy, and the consequences of the existence of such strategies. Alternatively and similarly, "determinacy" is the property of a game whereby such a strategy exists.

The games studied in set theory are usually Gale–Stewart games – two-player games of perfect information in which the players make an infinite sequence of moves and there are no draws. The field of game theory studies more general kinds of games, including games with draws such as tic-tac-toe, chess, or infinite chess, or games with imperfect information such as poker.

Basic notions

Games

The first sort of game we shall consider is the two-player game of perfect information of length ù, in which the players play natural numbers. These games are often called Gale–Stewart games.

In this sort of game there are two players, often named I and II, who take turns playing natural numbers, with I going first. They play "forever"; that is, their plays are indexed by the natural numbers. When they're finished, a predetermined condition decides which player won. This condition need not be specified by any definable rule; it may simply be an arbitrary (infinitely long) lookup table saying who has won given a particular sequence of plays.

More formally, consider a subset A of Baire space; recall that the latter consists of all ù-sequences of natural numbers. Then in the game GA, I plays a natural number a_0, then II plays a1, then I plays a2, and so on. Then I wins the game if and only if $(a_0, a_1, a_2,...)\in A$ and otherwise II wins. A is then called the payoff set of GA. It is assumed that each player can see all moves preceding each of his moves, and also knows the winning condition.

Strategies

Informally, a strategy for a player is a way of playing in which his plays are entirely determined by the foregoing plays. Again, such a "way" does not have to be capable of being captured by any explicable "rule", but may simply be a lookup table.

More formally, a strategy for player I (for a game in the sense of the preceding subsection) is a function that accepts as an argument any finite sequence of natural numbers, of even length, and returns a natural number. If ó is such a strategy and <a0,…,a2n-1> is a sequence of plays, then ó(<a0,…,a2n-1>) is the next play I will make, if he is following the strategy ó. Strategies for II are just the same, substituting "odd" for "even".

Note that we have said nothing, as yet, about whether a strategy is in any way good. A strategy might direct a player to make aggressively bad moves, and it would still be a strategy. In fact it is not necessary even to know the winning condition for a game, to know what strategies exist for the game.

Forcing

Paul Cohen invented the method of forcing while searching for a model of ZFC in which the continuum hypothesis fails, or a model of ZF in which the axiom of choice fails. Forcing adjoins to some given model of set theory additional sets in order to create a larger model with properties determined (i.e. "forced") by the construction and the original model. For example, Cohen's construction adjoins additional subsets of the natural numbers without changing any of the cardinal numbers of the original model. Forcing is also one of two methods for proving relative consistency by finitistic methods, the other method being Boolean-valued models.

Forcing (Mathematics)

In the mathematical discipline of set theory, forcing is a technique for proving consistency and independence results. It was first used by Paul Cohen in 1963, to prove the independence of the axiom of choice and the continuum hypothesis from Zermelo–Fraenkel set theory. Forcing has been considerably reworked and simplified in the following years, and has since served as a powerful technique, both in set theory and in areas of mathematical logic such as recursion theory. Descriptive set theory uses the notion of forcing from both recursion theory and set theory. Forcing has also been used in model theory, but it is common in model theory to define genericity directly without mention of forcing.

Intuition

Intuitively, forcing consists of expanding the set theoretical universe to a larger universe. In this bigger universe, for example, one might have many new subsets of that were not there in the old universe, and thereby violate the continuum hypothesis.

While impossible when dealing with finite sets, this is just another version of Cantor's paradox about infinity. In principle, one could consider: V " = V × {0, 1}, x∈V with (x,0) and then introduce an expanded membership relation involving "new" sets of the form (x, 1). Forcing is a more elaborate version of this idea, reducing the expansion to the existence of one new set, and allowing for fine control over the properties of the expanded universe. Cohen's original technique, now called ramified forcing, is slightly different from the unramified forcing expounded here. Forcing is also equivalent to the method of Boolean-valued Boolean-valued models, which some feel is conceptually more natural and intuitive, but usually much more difficult to apply.

Cardinal Invariants

A cardinal invariant is a property of the real line measured by a cardinal number. For example, a well-studied invariant is the smallest cardinality of a collection of meagre sets of reals whose union is the entire real line. These are invariants in the sense that any two isomorphic models of set theory must give the same cardinal for each invariant. Many cardinal invariants have been studied, and the relationships between them are often complex and related to axioms of set theory.

Cardinal Function

In mathematics, a cardinal function (or cardinal invariant) is a function that returns cardinal numbers.

Cardinal Functions in set Theory

- The most frequently used cardinal function is a function which assigns to a set "A" its cardinality, denoted by | A
- Aleph numbers and beth numbers can both be seen as cardinal functions defined on ordinal numbers.
- Cardinal arithmetic operations are examples of functions from cardinal numbers (or pairs of them) to cardinal numbers.
- Cardinal characteristics of a (proper) ideal I of subsets of X are:

The "additivity" of I is the smallest number of sets from I whose union is not in I any more. As any ideal is closed under finite unions, this number is always at least ; if I is a σ-ideal, then The "covering number" of I is the smallest number of sets from I whose union is all of X. As X itself is not in I, we must have add(I) dH cov(I).The "uniformity number" of I (sometimes also written u n i f (I) is the size of the smallest set not in I. Assuming I contains all singletons, add(I) dH non(I).The "cofinality" of I is the cofinality of the partial order (I, †"). It is easy to see that we must have non(I) d" cof(I) and cov(I) d" cof(I).

In the case that is an ideal closely related to the structure of the reals, such as the ideal of Lebesgue null sets or the ideal of meagre sets, these cardinal invariants are referred to as cardinal characteristics of the continuum.

- For a preordered set the bounding number and dominating number is defined as
- In PCF theory the cardinal function is used.

Cardinal functions are widely used in topology as a tool for describing various topological properties. Below are some examples. (Note: some authors, arguing that "there are no finite cardinal numbers in general topology", prefer to define the cardinal functions listed below so that they never taken on finite cardinal numbers as values; this requires modifying some of the definitions given below, e.g. by adding "" to the right-hand side of the definitions, etc.)

- Perhaps the simplest cardinal invariants of a topological space X are its cardinality and the cardinality of its topology, denoted respectively by |X| and o(X).
- The weight w(X) of a topological space X is the cardinality of the smallest base for X. When w(X) = the space X is said to be second countable.
- The -weight of a space X is the cardinality of the smallest -base for X.
- The character of a topological space X at a point x is the cardinality of the smallest local base for x. The character of space X is

When the space X is said to be first countable.

- The density d(X) of a space X is the cardinality of the smallest dense subset of X. When the space X is said to be separable.
- The Lindelöf number L(X) of a space X is the smallest infinite cardinality such that every open cover has a subcover of cardinality no more than L(X). When the space X is said to be a Lindelöf space.
- The cellularity or Suslin number of a space X is a family of mutually disjoint non-empty open subsets of.
- The hereditary cellularity (sometimes spread) is the least upper bound of cellularities of its subsets:

or

with the subspace topology is discrete

- The tightness t(x, X) of a topological space X at a point is the smallest cardinal number such that, whenever for some subset Y of X, there exists a subset Z of Y, with |Z| <, such that. Symbolically, The tightness of a space X is. When t(X) = the space X is said to be countably generated or countably tight.
- The augmented tightness of a space X, is the smallest regular cardinal such that for any, there is a subset Z of Y with cardinality less than, such that.

Basic Inequalities

c(X) < d(X)<w(X) ? o(X)< 2|X|

(X) < w(X)

nw(X) < w(X) and o(X) < 2nw(X)

Cardinal Functions in Boolean algebras

Cardinal functions are often used in the study of Boolean algebras. We can mention, for example, the following functions:

- Cellularity of a Boolean algebra is the supremum of the cardinalities of antichains in.
- Length of a Boolean algebra is
- is a chain
- Depth of a Boolean algebra is is a well-ordered subset.
- Incomparability of a Boolean algebra is
- such that.
- Pseudo-weight of a Boolean algebra is such that

Cardinal functions in algebra

Examples of cardinal functions in algebra are:

- Index of a subgroup H of G is the number of cosets.
- Dimension of a vector space V over a field K is the cardinality of any Hamel basis of V.
- More generally, for a free module M over a ring R we define rank as the cardinality of any basis of this module.
- For a linear subspace W of a vector space V we define codimension of W (with respect to V).
- For any algebraic structure it is possible to consider the minimal cardinality of generators of the structure.
- For algebraic extensions algebraic degree and separable degree are often employed (note that the algebraic degree equals the dimension of the extension as a vector space over the smaller field).
- For non-algebraic field extensions transcendence degree is likewise used.

Set-Theoretic Topology

Set-theoretic topology studies questions of general topology that are set-theoretic in nature or that require advanced methods of set theory for their solution. Many of these theorems are independent of ZFC, requiring stronger axioms for their proof. A famous problem is the normal Moore space question, a question in general topology that was the subject of intense research. The answer to the normal Moore space question was eventually proved to be independent of ZFC.

Set-theoretic topology

The space of integers has cardinality, while the real numbers has cardinality. The topologies of both spaces have cardinality. These are examples of cardinal functions, a topic in set-theoretic topology. In mathematics, set-theoretic topology is a subject that combines set theory and general topology. It focuses on topological questions that are independent of Zermelo-Fraenkel set theory (ZFC).

OBJECTS STUDIED IN SET-THEORETIC TOPOLOGY

Dowker Spaces

In the mathematical field of general topology, a Dowker space is a topological space that is T4 but not countably paracompact. Dowker conjectured that there were no Dowker spaces, and the conjecture was not resolved until M.E. Rudin constructed one in 1971. Rudin's counterexample is a very large space (of cardinality) and is generally not well-behaved. Zoltán Balogh gave the first ZFC construction of a small (cardinality continuum) example, which was more well-behaved than Rudin's. Using PCF theory, M. Kojman and S. Shelah constructed a subspace of Rudin's Dowker space of cardinality that is also Dowker.

Normal Moore Spaces

A famous problem is the normal Moore space question, a question in general topology that was the subject of intense research. The answer to the normal Moore space question was eventually proved to be independent of ZFC.

Cardinal Functions

Cardinal functions are widely used in topology as a tool for describing various topological properties. Below are some examples. (Note: some authors, arguing that "there are no finite cardinal numbers in general topology", prefer to define the cardinal functions listed below so that they never take on finite cardinal numbers as values; this requires modifying some of the definitions given below, e.g. by adding "" to the right-hand side of the definitions, etc.)

- Perhaps the simplest cardinal invariants of a topological space X are its cardinality and the cardinality of its topology, denoted respectively by |X| and o(X).
- The weight w (X) of a topological space X is the smallest possible cardinality of a base for X. When w (X) the space X is said to be second countable.
- The -weight of a space X is the smallest cardinality of a -base for X.
- The character of a topological space X at a point x is the smallest cardinality of a local base for x. The character of space X is When the space X is said to be first countable.
- The density d(X?) of a space X is the smallest cardinality of a dense subset of X. When the space X is said to be separable.
- The Lindelöf number L(X?) of a space X is the smallest infinite cardinality such that every open cover has a subcover of cardinality no more than L(X?). When the space X is said to be a Lindelöf space.
- The cellularity of a space X is is a family of mutually disjoint non-empty open subsets of.
- The Hereditary cellularity (sometimes spread) is the least upper bound of cellularities of its subsets:

  or

- The tightness t(x, X) of a topological space X at a point is the smallest cardinal number such that, whenever for some subset Y of X, there exists a subset Z of Y, with |Z|<, such that. Symbolically, The tightness of a space X is. When t(X) = the space X is said to be countably generated or countably tight.
- The augmented tightness of a space X, is the smallest regular cardinal such that for any, there is a subset Z of Y with cardinality less than, such that.

Martin's Axiom

For any cardinal k, we define a statement, denoted by MA(k): For any partial order P satisfying the countable chain condition (hereafter ccc) and any family D of dense sets in P such that |D| dH k, there is a filter F on P such that F)H d is non-empty for every d in D.Since it is a theorem of ZFC that MA(c) fails, the Martin's axiom is stated as:

Martin's axiom (MA): For every k < c, MA(k) holds.

In this case (for application of ccc), an antichain is a subset A of P such that any two distinct members of A are incompatible (two elements are said to be compatible if there exists a common element below both of them in the partial order). This differs from, for example, the notion of antichain in the context of trees.

MA() is false: [0, 1] is a compact Hausdorff space, which is separable and so ccc. It has no isolated points, so points in it are nowhere dense, but it is the union of many points.An equivalent formulation is: If X is a compact Hausdorff topological space which satisfies the ccc then X is not the union of k or fewer nowhere dense subsets. Martin's axiom has a number of other interesting combinatorial, analytic and topological consequences:

An equivalent formulation is: If X is a compact Hausdorff topological space which satisfies the ccc then X is not the union of k or fewer nowhere dense subsets.

Martin's axiom has a number of other interesting combinatorial, analytic and topological consequences:

- The union of k or fewer null sets in an atomless ó-finite Borel measure on a Polish space is null. In particular, the union of k or fewer subsets of R of Lebesgue measure 0 also has Lebesgue measure 0.
- A compact Hausdorff space X with $|X| < 2k$ is sequentially compact, i.e., every sequence has a convergent subsequence.
- No non-principal ultrafilter on N has a base of cardinality < k.
- Equivalently for any x in N\N we have $\chi(x) \geq k$, where χ is the character of x, and so $\chi(bx) \geq k$
- MA() implies that a product of ccc topological spaces is ccc (this in turn implies there are no Suslin lines).
- MA + ¬CH implies that there exists a Whitehead group that is not free; Shelah used this to show that the Whitehead problem is independent of ZFC.

Forcing

Forcing is a technique invented by Paul Cohen for proving consistency and independence results. It was first used, in 1963, to prove the independence of the axiom of choice and the continuum hypothesis from Zermelo–Fraenkel set theory. Forcing was considerably reworked and simplified in the 1960s, and has proven to be an extremely powerful technique both within set theory and in areas of mathematical logic such as recursion theory.

Intuitively, forcing consists of expanding the set theoretical universe V to a larger universe V*. In this bigger universe, for example, one might have lots of new subsets of ù = {0,1,2,...} that were not there in the old universe, and thereby violate the continuum hypothesis. While impossible on the face of it, this is just another version of Cantor's paradox about infinity. In principle, one could consider identify with, and then introduce an expanded membership relation involving the "new" sets of the form. Forcing is a more elaborate version of this idea, reducing the expansion to the existence of one new set, and allowing for fine control over the properties of the expanded universe.

OBJECTIONS TO SET THEORY AS A FOUNDATION FOR MATHEMATICS

From set theory's inception, some mathematicians have objected to it as a foundation for mathematics. The most common objection to set theory, one Kronecker voiced in set theory's earliest years, starts from the constructivist view that mathematics is loosely related to computation. If this view is granted, then the treatment of infinite sets, both in naive and in axiomatic set theory, introduces into mathematics methods and objects that are not computable even in principle. The feasibility of constructivism as a substitute foundation for mathematics was greatly increased by Errett Bishop's influential book Foundations of Constructive Analysis.

A different objection put forth by Henri Poincaré is that defining sets using the axiom schemas of specification and replacement, as well as the axiom of power set, introduces impredicativity, a type of circularity, into the definitions of mathematical objects. The scope of predicatively founded mathematics, while less than that of the commonly accepted Zermelo-Fraenkel theory, is much greater than that of constructive mathematics, to the point that Solomon Feferman has said that "all of scientifically applicable analysis can be developed [using predicative methods]".

Ludwig Wittgenstein condemned set theory. He wrote that "set theory is wrong", since it builds on the "nonsense" of fictitious symbolism, has "pernicious idioms", and that it is nonsensical to talk about "all numbers". Wittgenstein's views about the foundations of mathematics were later criticised by Georg Kreisel and Paul Bernays, and investigated by Crispin Wright, among others. Category theorists have proposed topos theory as an alternative to traditional axiomatic set theory. Topos theory can interpret various alternatives to that theory, such as constructivism, finite set theory, and computable set theory. Topoi also give a natural setting for forcing and discussions of the independence of choice from ZF, as well as providing the framework for pointless topology and Stone spaces.

An active area of research is the univalent foundations and related to it homotopy type theory. Within homotopy type theory, a set may be regarded as a homotopy 0-type, with universal properties of sets arising from the inductive and recursive properties of higher inductive types. Principles such as the axiom of choice and the law of the excluded middle can be formulated in a manner corresponding to the classical formulation in set theory or perhaps in a spectrum of distinct ways unique to type theory. Some of these principles may be proven to be a consequence of other principles. The variety of formulations of these axiomatic principles allows for a detailed analysis of the formulations required in order to derive various mathematical results.

5

COMMUTATIVE ALGEBRA APPLICATIONS

Commutative algebra, first known as ideal theory, is the branch of algebra that studies commutative rings, their ideals, and modules over such rings. Both algebraic geometry and algebraic number theory build on commutative algebra. Prominent examples of commutative rings include polynomial rings; rings of algebraic integers, including the ordinary integers Z {mathbb {Z}} \mathbb {Z} ; and p-adic integers.

Commutative algebra is the main technical tool in the local study of schemes. The term "commutative algebra" also refers to the branch of abstract algebra that studies commutative rings. Commutative algebra is important in algebraic geometry.

The study of rings that are not necessarily commutative is known as noncommutative algebra; it includes ring theory, representation theory, and the theory of Banach algebras. Commutative algebra is essentially the study of the rings occurring in algebraic number theory and algebraic geometry.

In algebraic number theory, the rings of algebraic integers are Dedekind rings, which constitute therefore an important class of commutative rings. Considerations related to modular arithmetic have led to the notion of a valuation ring. The restriction of algebraic field extensions to subrings has led to the notions of integral extensions and integrally closed domains as well as the notion of ramification of an extension of valuation rings.

The notion of localization of a ring (in particular the localization with respect to a prime ideal, the localization consisting in inverting a single element and the total quotient ring) is one of the main differences between commutative algebra and the theory of non-commutative rings. It leads to an important class of commutative rings, the local rings that have only one maximal ideal. The set of the prime ideals of a commutative ring is naturally equipped with a topology, the Zariski topology. All these notions are widely used in algebraic geometry and are the basic technical tools for the definition of scheme theory, a generalization of algebraic geometry introduced by Grothendieck.

Many other notions of commutative algebra are counterparts of geometrical notions occurring in algebraic geometry. This is the case of Krull dimension, primary decomposition, regular rings, Cohen–Macaulay rings, Gorenstein rings and many other notions.

Localization and completion are among the fundamental first tools in commutative algebra. They play a correspondingly fundamental role in algebraic topology. Localizations and completions of spaces and spectra have been central tools since the 1970's. Some basic references are,,. These constructions start from ideals in the ring of integers and are very simple algebraically since Z is a principal ideal domain. Localizations and completions that start from ideals in the representation ring or the Burnside ring of a compact Lie group play a correspondingly central role in equivariant topology.

These rings are still relatively simple algebraically since, when G is finite, they are Noetherian and of Krull dimension one. A common general framework starts from ideals in the coefficient ring of a generalized cohomology theory. We shall explain some old and new algebra that arises in this context, and we will show how this algebra can be mimicked topologically. The topological constructions require the foundations described in the previous article, which deals with the algebraically familiar theory of localization at multiplicatively closed subsets. We here explain the deeper and less familiar theory of completion, together with an ideal theoretic variant of localization. There is a still more general theory of localization of spaces and spectra at spectra, due to Bousfield,, and we shall see how our theory of localizations and completions with respect to ideals in coefficient rings fits into this context.

HISTORY

The subject, first known as ideal theory, began with Richard Dedekind's work on ideals, itself based on the earlier work of Ernst Kummer and Leopold Kronecker. Later, David Hilbert introduced the term ring to generalize the earlier term number ring. Hilbert introduced a more abstract approach to replace the more concrete and computationally oriented methods grounded in such things as complex analysis and classical invariant theory. In turn, Hilbert strongly influenced Emmy Noether, who recast many earlier results in terms of an ascending chain condition, now known as the Noetherian condition. Another important milestone was the work of Hilbert's student Emanuel Lasker, who introduced primary ideals and proved the first version of the Lasker–Noether theorem.

The main figure responsible for the birth of commutative algebra as a mature subject was Wolfgang Krull, who introduced the fundamental notions of localization and completion of a ring, as well as that of regular local rings. He established the concept of the Krull dimension of a ring, first for Noetherian rings before moving on to expand his theory to cover general valuation rings and Krull rings. To this day, Krull's principal ideal theorem is widely considered the single most important foundational theorem in commutative algebra. These results paved the way for the introduction of commutative algebra into algebraic geometry, an idea which would revolutionize the latter subject. Much of the modern development of commutative algebra emphasizes modules. Both ideals of a ring R and R-algebras are special cases of R-modules, so module theory encompasses both ideal theory and the theory of ring extensions. Though it was already incipient in Kronecker's work, the modern approach to commutative algebra using module theory is usually credited to Krull and Noether.

Noetherian rings

In mathematics, more specifically in the area of modern algebra known as ring theory, a Noetherian ring, named after Emmy Noether, is a ring in which every non-empty set of ideals has a maximal element. Equivalently, a ring is Noetherian if it satisfies the ascending chain condition on ideals; that is, given any chain: $I_1 \subseteq \cdots I_{k-1} \subseteq I_k \subseteq I_{k+1} \subseteq \cdots$there exists an n such that: $I_n = I_{n+1} = \cdots$

For a commutative ring to be Noetherian it suffices that every prime ideal of the ring is finitely generated. (The result is due to I. S. Cohen.) The notion of a Noetherian ring is of fundamental importance in both commutative and noncommutative ring theory, due to the role it plays in simplifying the ideal structure of a ring. For instance, the ring of integers and the polynomial ring over a field are both Noetherian rings, and consequently, such theorems as the Lasker–Noether theorem, the Krull intersection theorem, and the Hilbert's basis theorem hold for them. Furthermore, if a ring is Noetherian, then it satisfies the descending chain condition on prime ideals. This property suggests a deep theory of dimension for Noetherian rings beginning with the notion of the Krull dimension.$I_1 \subseteq \cdots I_2 \subseteq I_3 \subseteq \cdots$

In mathematics, a Noetherian ring is a ring that satisfies the ascending chain condition on left and right ideals; if the chain condition is satisfied only for left ideals or for right ideals, then the ring is said left-Noetherian or right-Noetherian respectively. That is, every increasing sequence of left (or right) ideals has a largest element; that is, there exists an n such that: $I_n = I_{n+1} = \cdots$.

Equivalently, a ring is left-Noetherian (resp. right-Noetherian) if every left ideal (resp. right-ideal) is finitely generated. A ring is Noetherian if it is both left- and right-Noetherian.

Noetherian rings are fundamental in both commutative and noncommutative ring theory since many rings that are encountered in mathematics are Noetherian (in particular the ring of integers, polynomial rings, and rings of algebraic integers), and many general theorems on rings rely heavily on Noetherian property (for example, the Lasker–Noether theorem and the Krull intersection theorem).

Noetherian rings are named after Emmy Noether, but the importance of the concept was recognized earlier by David Hilbert, with the proof of Hilbert's basis theorem (which asserts that polynomial rings are Noetherian) and Hilbert's syzygy theorem.

Properties

If R is a Noetherian ring, then the polynomial ring R [X]]is Noetherian by the Hilbert's basis theorem. By induction, R [X 1, …, X n] is a Noetherian ring. Also, R[[X]], the power series ring, is a Noetherian ring.

If R is a Noetherian ring and I is a two-sided ideal, then the quotient ring R/I is also Noetherian. Stated differently, the image of any surjective ring homomorphism of a Noetherian ring is Noetherian.

Every finitely-generated commutative algebra over a commutative Noetherian ring is Noetherian. (This follows from the two previous properties.)

A ring R is left-Noetherian if and only if every finitely generated left R-module is a Noetherian module.

If a commutative ring admits a faithful Noetherian module over it, then the ring is a Noetherian ring.

(Eakin–Nagata) If a ring A is a subring of a commutative Noetherian ring B such that B is a finitely generated module over A, then A is a Noetherian ring.

Similarly, if a ring A is a subring of a commutative Noetherian ring B such that B is faithfully flat over A (or more generally exhibits A as a pure subring), then A is a Noetherian ring (see the "faithfully flat" article for the reasoning).

Every localization of a commutative Noetherian ring is Noetherian.

A consequence of the Akizuki–Hopkins–Levitzki theorem is that every left Artinian ring is left Noetherian. Another consequence is that a left Artinian ring is right Noetherian if and only if it is right Artinian. The analogous statements with "right" and "left" interchanged are also true.

A left Noetherian ring is left coherent and a left Noetherian domain is a left Ore domain.

(Bass) A ring is (left/right) Noetherian if and only if every direct sum of injective (left/right) modules is injective. Every left injective module over a left Noetherian module can be decomposed as a direct sum of indecomposable injective modules. See also #Implication on injective modules below.

In a commutative Noetherian ring, there are only finitely many minimal prime ideals. Also, the descending chain condition holds on prime ideals.

In a commutative Noetherian domain R, every element can be factorized into irreducible elements (in short, R is a factorization domain). Thus, if, in addition, the factorization is unique up to multiplication of the factors by units, then R is a unique factorization domain.

Primary decomposition

An ideal Q of a ring is said to be primary if Q is proper and whenever xy " Q, either x " Q or yn " Q for some positive integer n. In Z, the primary ideals are precisely the ideals of the form (pe) where p is prime and e is a positive integer. Thus, a primary decomposition of (n) corresponds to representing (n) as the intersection of finitely many primary ideals. The Lasker–Noether theorem, given here, may be seen as a certain generalization of the fundamental theorem of arithmetic: Lasker-Noether Theorem. Let R be a commutative Noetherian ring and let I be an ideal of R. Then I may be written as the intersection of finitely many primary ideals with distinct radicals; that is: $I = \bigcap_{i=1}^{t} Q_i$

with Qi primary for all i and Rad(Qi) "' Rad(Qj) for i "' j. Furthermore, if: $I = \bigcap_{i=1}^{k} P_i$

is decomposition of I with Rad(Pi) "' Rad(Pj) for i "' j, and both decompositions of I are irredundant (meaning that no proper subset of either {Q1,..., Qt} or {P1,..., Pk} yields an intersection equal to I), t = k and (after possibly renumbering the Qi) Rad(Qi) = Rad(Pi) for all i. For any primary decomposition of I, the set of all radicals, that is, the set {Rad(Q1),..., Rad(Qt)} remains the same by the Lasker–Noether theorem. In fact, it turns out that (for a Noetherian ring) the set is precisely the assassinator of the module R/I; that is, the set of all annihilators of R/I (viewed as a module over R) that are prime.

Localization

The localization is a formal way to introduce the "denominators" to a given ring or a module. That is, it introduces a new ring/module out of an existing one so that it consists of fractions $\frac{m}{s}$

where the denominators s range in a given subset S of R. The archetypal example is the construction of the ring Q of rational numbers from the ring Z of integers.

Completion

A completion is any of several related functors on rings and modules that result in complete topological rings and modules. Completion is similar to localization, and together they are among the most basic tools in analysing commutative rings. Complete commutative rings have simpler structure than the general ones and Hensel's lemma applies to them. Zariski topology on prime ideals The Zariski topology defines a topology on the spectrum of a ring (the set of prime ideals). In this formulation, the Zariski-closed sets are taken to be the sets $V(I) = \{P \in \text{Spec}(A) | I \subseteq P\}$ where A is a fixed commutative ring and I is an ideal. This is defined in analogy with the classical Zariski topology, where closed sets in affine space are those defined by polynomial equations. To see the connection with the classical picture, note that for any set S of polynomials (over an algebraically closed field), it follows from Hilbert's Nullstellensatz that the points of V(S) (in the old sense) are exactly the tuples (a1,..., an) such that (x1 - a1,..., xn - an) contains S; moreover, these are maximal ideals and by the "weak" Nullstellensatz, an ideal of any affine coordinate ring is maximal if and only if it is of this form. Thus, V(S) is "the same as" the maximal ideals containing S. Grothendieck's innovation in defining Spec was to replace maximal ideals with all prime ideals; in this formulation it is natural to simply generalize this observation to the definition of a closed set in the spectrum of a ring.

Examples

The fundamental example in commutative algebra is the ring of integers Z. The existence of primes and the unique factorization theorem laid the foundations for concepts such as Noetherian rings and the primary decomposition.

Other important examples are:

Polynomial rings

- The p-adic integers
- Rings of algebraic integers.

CNNECTIONS WITH ALGEBRAIC GEOMETRY

Commutative algebra (in the form of polynomial rings and their quotients, used in the definition of algebraic varieties) has always been a part of algebraic geometry. However, in the late 1950s, algebraic varieties were subsumed into Alexander Grothendieck's concept of a scheme.

Their local objects are affine schemes or prime spectra, which are locally ringed spaces, which form a category that is antiequivalent (dual) to the category of commutative unital rings, extending the duality between the category of affine algebraic varieties over a field k, and the category of finitely generated reduced k-algebras.

The gluing is along the Zariski topology; one can glue within the category of locally ringed spaces, but also, using the Yoneda embedding, within the more abstract category of presheaves of sets over the category of affine schemes. The Zariski topology in the set-theoretic sense is then replaced by a Zariski topology in the sense of Grothendieck topology.

Grothendieck introduced Grothendieck topologies having in mind more exotic but geometrically finer and more sensitive examples than the crude Zariski topology, namely the étale topology, and the two flat Grothendieck topologies: fppf and fpqc. Nowadays some other examples have become prominent, including the Nisnevich topology. Sheaves can be furthermore generalized to stacks in the sense of Grothendieck, usually with some additional representability conditions, leading to Artin stacks and, even finer, Deligne-Mumford stacks, both often called algebraic stacks.

Combinatorial commutative algebra

Combinatorial commutative algebra is a relatively new, rapidly developing mathematical discipline. As the name implies, it lies at the intersection of two more established fields, commutative algebra and combinatorics, and frequently uses methods of one to address problems arising in the other. Less obviously, polyhedral geometry plays a significant role.

One of the milestones in the development of the subject was Richard Stanley's 1975 proof of the Upper Bound Conjecture for simplicial spheres, which was based on earlier work of Melvin Hochster and Gerald Reisner. While the problem can be formulated purely in geometric terms, the methods of the proof drew on commutative algebra techniques.

A signature theorem in combinatorial commutative algebra is the characterization of h-vectors of simplicial polytopes conjectured in 1970 by Peter McMullen. Known as the g-theorem, it was proved in 1979 by Stanley (necessity of the conditions, algebraic argument) and by Louis Billera and Carl W. Lee (sufficiency, combinatorial and geometric construction). A major open question was the extension of this characterization from simplicial polytopes to simplicial spheres, the g-conjecture, which was resolved in 2018 by Karim Adiprasito.

6

GROUP THEORY : DEFINITION AND APPLICATIONS

In abstract algebra, group theory studies the algebraic structures known as groups. The concept of a group is central to abstract algebra: other well-known algebraic structures, such as rings, fields, and vector spaces, can all be seen as groups endowed with additional operations and axioms. Groups recur throughout mathematics, and the methods of group theory have influenced many parts of algebra. Linear algebraic groups and Lie groups are two branches of group theory that have experienced advances and have become subject areas in their own right.

Various physical systems, such as crystals and the hydrogen atom, and three of the four known fundamental forces in the universe, may be modelled by symmetry groups. Thus group theory and the closely related representation theory have many important applications in physics, chemistry, and materials science. Group theory is also central to public key cryptography.

The early history of group theory dates from the 19th century. One of the most important mathematical achievements of the 20th century was the collaborative effort, taking up more than 10,000 journal pages and mostly published between 1960 and 2004, that culminated in a complete classification of finite simple groups.

HISTORY OF GROUP THEORY

The modern concept of an abstract group developed out of several fields of mathematics. The original motivation for group theory was the quest for solutions of polynomial equations of degree higher than 4. The 19th-century French mathematician Évariste Galois, extending prior work of Paolo Ruffini and Joseph-Louis Lagrange, gave a criterion for the solvability of a particular polynomial equation in terms of the symmetry group of its roots (solutions). The elements of such a Galois group correspond to certain permutations of the roots. At first, Galois' ideas were rejected by his contemporaries, and published only posthumously. More general permutation groups were investigated in particular by Augustin Louis Cauchy. Arthur Cayley's On the theory of groups, as depending on the symbolic equation èn = 1 (1854) gives the first abstract definition of a finite group. Geometry was a second field in which groups were used systematically, especially symmetry groups as part of Felix Klein's 1872 Erlangen program. After novel geometries such as hyperbolic and projective geometry had emerged, Klein used group theory to organize them in a more coherent way. Further advancing these ideas, Sophus Lie founded the study of Lie groups in 1884. The third field contributing to group theory was number theory. Certain abelian group structures had been used implicitly in Carl Friedrich Gauss' number-theoretical work Disquisitiones Arithmeticae (1798), and more explicitly by Leopold Kronecker. In 1847, Ernst Kummer made early attempts to prove Fermat's Last Theorem by developing groups describing factorization into prime numbers.

Group theory has three main historical sources: number theory, the theory of algebraic equations, and geometry. The number-theoretic strand was begun by Leonhard Euler, and developed by Gauss's work on modular arithmetic and additive and multiplicative groups related to quadratic fields. Early results about permutation groups were obtained by Lagrange, Ruffini, and Abel in their quest for general solutions of polynomial equations of high degree. Évariste Galois coined the term "group" and established a connection, now known as Galois theory, between the nascent theory of groups and field theory. In geometry, groups first became important in projective geometry and, later, non-Euclidean geometry. Felix Klein's Erlangen program proclaimed group theory to be the organizing principle of geometry.

Galois, in the 1830s, was the first to employ groups to determine the solvability of polynomial equations. Arthur Cayley and Augustin Louis Cauchy pushed these investigations further by creating the theory of permutation groups. The second historical source for groups stems from geometrical situations. In an attempt to come to grips with possible geometries (such as euclidean, hyperbolic or projective geometry) using group theory, Felix Klein initiated the Erlangen programme. Sophus Lie, in 1884, started using groups (now called Lie groups) attached to analytic problems. Thirdly, groups were, at first implicitly and later explicitly, used in algebraic number theory.

The different scope of these early sources resulted in different notions of groups. The theory of groups was unified starting around 1880. Since then, the impact of group theory has been ever growing, giving rise to the birth of abstract algebra in the early 20th century, representation theory, and many more influential spin-off domains. The classification of finite simple groups is a vast body of work from the mid 20th century, classifying all the finite simple groups.

The convergence of these various sources into a uniform theory of groups started with Camille Jordan's Traité des substitutions et des équations algébriques (1870). Walther von Dyck (1882) introduced the idea of specifying a group by means of generators and relations, and was also the first to give an axiomatic definition of an "abstract group", in the terminology of the time. As of the 20th century, groups gained wide recognition by the pioneering work of Ferdinand Georg Frobenius and William Burnside, who worked on representation theory of finite groups, Richard Brauer's modular representation theory and Issai Schur's papers. The theory of Lie groups, and more generally locally compact groups was studied by Hermann Weyl, Élie Cartan and many others. Its algebraic counterpart, the theory of algebraic groups, was first shaped by Claude Chevalley (from the late 1930s) and later by the work of Armand Borel and Jacques Tits.

The University of Chicago's 1960–61 Group Theory Year brought together group theorists such as Daniel Gorenstein, John G. Thompson and Walter Feit, laying the foundation of a collaboration that, with input from numerous other mathematicians, led to the classification of finite simple groups, with the final step taken by Aschbacher and Smith in 2004. This project exceeded previous mathematical endeavours by its sheer size, in both length of proof and number of researchers. Research is ongoing to simplify the proof of this classification. These days, group theory is still a highly active mathematical branch, impacting many other fields.a[›

MAIN CLASSES OF GROUPS

The range of groups being considered has gradually expanded from finite permutation groups and special examples of matrix groups to abstract groups that may be specified through a presentation by generators and relations.

Permutation groups

The first class of groups to undergo a systematic study was permutation groups. Given any set X and a collection G of bijections of X into itself (known as permutations) that is closed under compositions and inverses, G is a group acting on X. If X consists of n elements and G consists of all permutations, G is the symmetric group Sn; in general, any permutation group G is a subgroup of the symmetric group of X. An early construction due to Cayley exhibited any group as a permutation group, acting on itself (X = G) by means of the left regular representation.

In many cases, the structure of a permutation group can be studied using the properties of its action on the corresponding set. For example, in this way one proves that for n = 5, the alternating group An is simple, i.e. does not admit any proper normal subgroups. This fact plays a key role in the impossibility of solving a general algebraic equation of degree n = 5 in radicals.

Matrix groups

The next important class of groups is given by matrix groups, or linear groups. Here G is a set consisting of invertible matrices of given order n over a field K that is closed under the products and inverses. Such a group acts on the n-dimensional vector space Kn by linear transformations. This action makes matrix groups conceptually similar to permutation groups, and the geometry of the action may be usefully exploited to establish properties of the group G.

Transformation groups

Permutation groups and matrix groups are special cases of transformation groups: groups that act on a certain space X preserving its inherent structure. In the case of permutation groups, X is a set; for matrix groups, X is a vector space. The concept of a transformation group is closely related with the concept of a symmetry group: transformation groups frequently consist of all transformations that preserve a certain structure.

The theory of transformation groups forms a bridge connecting group theory with differential geometry. A long line of research, originating with Lie and Klein, considers group actions on manifolds by homeomorphisms or diffeomorphisms. The groups themselves may be discrete or continuous.

Abstract groups

Most groups considered in the first stage of the development of group theory were "concrete", having been realized through numbers, permutations, or matrices. It was not until the late nineteenth century that the idea of an abstract group as a set with operations satisfying a certain system of axioms began to take hold. A typical way of specifying an abstract group is through a presentation by generators and relations, G = {S|R}.

A significant source of abstract groups is given by the construction of a factor group, or quotient group, G/H, of a group G by a normal subgroup H. Class groups of algebraic number fields were among the earliest examples of factor groups, of much interest in number theory. If a group G is a permutation group on a set X, the factor group G/H is no longer acting on X; but the idea of an abstract group permits one not to worry about this discrepancy.

The change of perspective from concrete to abstract groups makes it natural to consider properties of groups that are independent of a particular realization, or in modern language, invariant under isomorphism, as well as the classes of group with a given such property: finite groups, periodic groups, simple groups, solvable groups, and so on.

Rather than exploring properties of an individual group, one seeks to establish results that apply to a whole class of groups. The new paradigm was of paramount importance for the development of mathematics: it foreshadowed the creation of abstract algebra in the works of Hilbert, Emil Artin, Emmy Noether, and mathematicians of their school.

Groups with additional structure]

An important elaboration of the concept of a group occurs if G is endowed with additional structure, notably, of a topological space, differentiable manifold, or algebraic variety. If the group operations m (multiplication) and i (inversion), $m : G\times G\to G,(g,h)\mapsto gh,\ i: G\to G, g\mapsto g^{-1}$

are compatible with this structure, that is, they are continuous, smooth or regular (in the sense of algebraic geometry) maps, then G is a topological group, a Lie group, or an algebraic group.

The presence of extra structure relates these types of groups with other mathematical disciplines and means that more tools are available in their study. Topological groups form a natural domain for abstract harmonic analysis, whereas Lie groups (frequently realized as transformation groups) are the mainstays of differential geometry and unitary representation theory. Certain classification questions that cannot be solved in general can be approached and resolved for special subclasses of groups. Thus, compact connected Lie groups have been completely classified. There is a fruitful relation between infinite abstract groups and topological groups: whenever a group G can be realized as a lattice in a topological group G, the geometry and analysis pertaining to G yield important results about G. A comparatively recent trend in the theory of finite groups exploits their connections with compact topological groups (profinite groups): for example, a single p-adic analytic group G has a family of quotients which are finite p-groups of various orders, and properties of G translate into the properties of its finite quotients.

Definition and Illustration

First example: the integers

One of the most familiar groups is the set of integers which consists of the numbers..., 4, 3, 2, 1, 0, 1, 2, 3, 4,..., together with addition.The following properties of integer addition serve as a model for the group axioms given in the definition below.

- For any two integers a andb, the suma + b is also an integer. That is, addition of integers always yields an integer. This property is known as closure under addition.
- For all integers a, b and c, (a + b) + c = a + (b + c). Expressed in words, adding a tob first, and then adding the result to c gives the same final result as adding a to the sum of b and c, a property known as associativity.
- If a is any integer, then 0 + a = a + 0 = a. Zero is called the identity element of addition because adding it to any integer returns the same integer.
- For every integer a, there is an integer b such that a + b = b + a = 0. The integer b is called the inverse element of the integer a and is denoted Ha.

The integers, together with the operation +, form a mathematical object belonging to a broad class sharing similar structural aspects. To appropriately understand these structures as a collective, the following definition is developed.

DEFINITION

The axioms for a group are short and natural... Yet somehow hidden behind these axioms is the monster simple group, a huge and extraordinary mathematical object, which appears to rely on numerous bizarre coincidences to exist. The axioms for groups give no obvious hint that anything like this exists.

Richard Borcherds in Mathematicians: An Outer View of the Inner World A group is a set, G, together with an operation • (called the group law of G) that combines any two elementsa andb to form another element, denoted a • b or ab. To qualify as a group, the set and operation, (G, •), must satisfy four requirements known as the group axioms:

Closure

For all a, b in G, the result of the operation, a • b, is also in G.b[›]

Associativity

For all a, b and c in G, (a • b) • c = a • (b • c).

Identity element

There exists an element e in G such that, for every element a in G, the equation e • a = a • e = a holds. Such an element is unique, and thus one speaks of the identity element.

Inverse element

For each a in G, there exists an element b in G, commonly denoted a"1 (or "a, if the operation is denoted "+"), such that a • b = b • a = e, where e is the identity element.

denoted "+"), such that a • b = b • a = e, where e is the identity element. The result of an operation may depend on the order of the operands. In other words, the result of combining element a with element b need not yield the same result as combining element b with element a; the equation

a • b = b • a

may not always be true. This equation always holds in the group of integers under addition, because a + b = b + a for any two integers (commutativity of addition). Groups for which the commutativity equation a • b = b • a always holds are called abelian groups (in honor of Niels Henrik Abel). The symmetry group described in the following section is an example of a group that is not abelian. The identity element of a group G is often written as 1 or 1G, a notation inherited from the multiplicative identity. If a group is abelian, then one may choose to denote the group operation by + and the identity element by 0; in that case, the group is called an additive group. The identity element can also be written as id.

The set G is called the underlying set of the group (G, •). Often the group's underlying set G is used as a short name for the group (G, •). Along the same lines, shorthand expressions such as "a subset of the group G" or "an element of group G" are used when what is actually meant is "a subset of the underlying set G of the group (G, •)" or "an element of the underlying set G of the group (G, •)". Usually, it is clear from the context whether a symbol like G refers to a group or to an underlying set.

An alternate (but equivalent) definition is to expand the structure of a group to define a group as a set equipped with three operations satisfying the same axioms as above, with the "there exists" part removed in the two last axioms, these operations being the group law, as above, which is a binary operation, the inverse operation, which is a unary operation and maps a to a H $1a^{-2}$, and the identity element, which is viewed as a 0-ary operation.

As this formulation of the definition avoids existential quantifiers, it is generally preferred for computing with groups and for computer-aided proofs. This formulation exhibits groups as a variety of universal algebra. It is also useful for talking of properties of the inverse operation, as needed for defining topological groups and group objects.

Second Example: a Symmetry Group

Two figures in the plane are congruent if one can be changed into the other using a combination of rotations, reflections, and translations. Any figure is congruent to itself. However, some figures are congruent to themselves in more than one way, and these extra congruences are called symmetries. A square has eight symmetries. These are: The elements of the symmetry group of the square (D4). Vertices are identified by color or number.

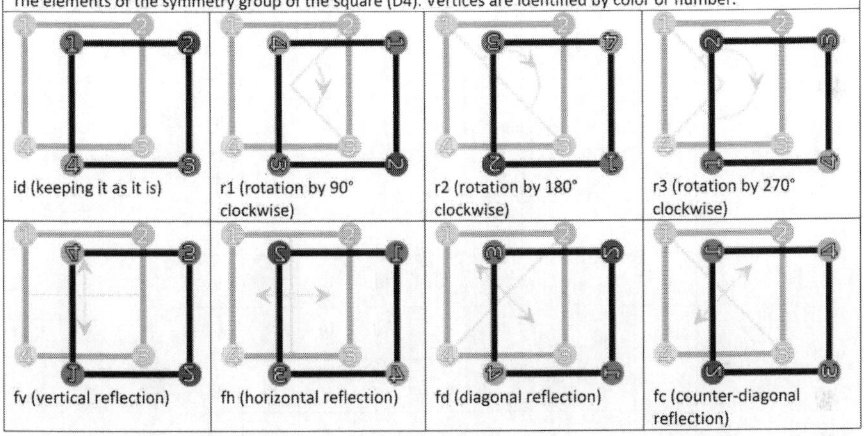

The elements of the symmetry group of the square (D4). Vertices are identified by color or number.

id (keeping it as it is); r1 (rotation by 90° clockwise); r2 (rotation by 180° clockwise); r3 (rotation by 270° clockwise); fv (vertical reflection); fh (horizontal reflection); fd (diagonal reflection); fc (counter-diagonal reflection)

- The identity operation leaving everything unchanged, denoted id;
- Rotations of the square around its center by 90° clockwise, 180° clockwise, and 270° clockwise, denoted by r1, r2 and r3, respectively;
- Reflections about the vertical and horizontal middle line (fh and fv), or through the two diagonals (fd and fc).

These symmetries are represented by functions. Each of these functions sends a point in the square to the corresponding point under the symmetry. For example, r1 sends a point to its rotation 90° clockwise around the square's center, and fh sends a point to its reflection across the square's vertical middle line. Composing two of these symmetry functions gives another symmetry function. These symmetries determine a group called the dihedral group of degree 4 and denoted D4. The underlying set of the group is the above set of symmetry functions, and the group operation is function composition. Two symmetries are combined by composing them as functions, that is, applying the first one to the square, and the second one to the result of the first application. The result of performing first a and then b is written symbolically from right to left as

b • a ("apply the symmetry b after performing the symmetry a").The right-to-left notation is the same notation that is used for composition of functions. The group table on the right lists the results of all such compositions possible. For example, rotating by 270° clockwise (r3) and then reflecting horizontally (fh) is the same as performing a reflection along the diagonal (fd). Using the above symbols, highlighted in blue in the group table:

fh • r3 = fd.

Group table of D_4

•	id	r_1	r_2	r_3	f_v	f_h	f_d	f_c
id	id	r_1	r_2	r_3	f_v	f_h	f_d	f_c
r_1	r_1	r_2	r_3	id	f_c	f_d	f_v	f_h
r_2	r_2	r_3	id	r_1	f_h	f_v	f_c	f_d
r_3	r_3	id	r_1	r_2	f_d	f_c	f_h	f_v
f_v	f_v	f_d	f_h	f_c	id	r_2	r_1	r_3
f_h	f_h	f_c	f_v	f_d	r_2	id	r_3	r_1
f_d	f_d	f_h	f_c	f_v	r_3	r_1	id	r_2
f_c	f_c	f_v	f_d	f_h	r_1	r_3	r_2	id

The elements id, r_1, r_2, and r_3 form a subgroup, highlighted in ☐ red (upper left region). A left and right coset of this subgroup is highlighted in ☐ green (in the last row) and ☐ yellow (last column), respectively.

Given this set of symmetries and the described operation, the group axioms can be understood as follows:

1. The closure axiom demands that the composition b • a of any two symmetries a and b is also a symmetry. Another example for the group operation is

r3 • fh = fc, i.e., rotating 270° clockwise after reflecting horizontally equals reflecting along the counter-diagonal (fc). Indeed every other combination of two symmetries still gives a symmetry, as can be checked using the group table.

2. The associativity constraint deals with composing more than two symmetries: Starting with three elements a, b and c of D4, there are two possible ways of using these three symmetries in this order to determine a symmetry of the square. One of these ways is to first compose a andb into a single symmetry, then to compose that symmetry with c. The other way is to first compose b and c, then to compose the resulting symmetry with a. The associativity condition (a • b) • c = a • (b • c)means that these two ways are the same, i.e., a product of many group elements can besimplified in any grouping. For example, (fd o fv) o r2 = fd o (fv o r2) can be checked using the group table at the right

$(f_d \bullet f_v) \bullet r_2$	=	$r_3 \bullet r_2$	=	r_1, which equals
$f_d \bullet (f_v \bullet r_2)$	=	$f_d \bullet f_h$	=	r_1.

not true for all operations. For instance, subtraction of numbers is not associative:

(7 “ 3) “ 2 = 2 is not the same as 7 “ (3 “ 2) = 6.

“ 2 = 2 is not the same as 7 “ (3 “ 2) = 6.

1. The identity element is the symmetry id leaving everything unchanged: for any symmetry a, performing id after a (or a after id) equals a, in symbolic form,

id • a = a,

a • id = a.

2. An inverse element undoes the transformation of some other element. Every symmetry can be undone: each of the following transformations—identity id, the reflections fh, fv, fd, fc and the 180° rotation r2—is its own inverse, because performing it twice brings the square back to its original orientation. The rotations r3 and r1 are each other's inverses, because rotating 90° and then rotation 270° (or vice versa) yields a rotation over 360° which leaves the square unchanged. In symbols,

fh • fh = id,

r3 • r1 = r1 • r3 = id.

In contrast to the group of integers above, where the order of the operation is irrelevant, it does matter in D4: fh • r1 = fc but r1 • fh = fd. In other words, D4 is not abelian, which makes the group structure more difficult than the integers introduced first.

Elementary consequences of the group axioms

Basic facts about all groups that can be obtained directly from the group axioms are commonly subsumed under elementary group theory. For example, repeated applications of the associativity axiom show that the unambiguity of

$$a \bullet b \bullet c = (a \bullet b) \bullet c = a \bullet (b \bullet c)$$

generalizes to more than three factors. Because this implies that parentheses can be inserted anywhere within such a series of terms, parentheses are usually omitted. The axioms may be weakened to assert only the existence of a left identity and left inverses. Both can be shown to be actually two-sided, so the resulting definition is equivalent to the one given above.

Uniqueness of Identity Element and Inverses

Two important consequences of the group axioms are the uniqueness of the identity element and the uniqueness of inverse elements. There can be only one identity element in a group, and each element in a group has exactly one inverse element. Thus, it is customary to speak of the identity, and the inverse of an element. To prove the uniqueness of an inverse element of a, suppose that a has two inverses, denoted b and c, in a group (G, •). Then

b	=	$b \bullet e$		as e is the identity element
	=	$b \bullet (a \bullet c)$		because c is an inverse of a, so $e = a \bullet c$
	=	$(b \bullet a) \bullet c$		by associativity, which allows rearranging the parentheses
	=	$e \bullet c$		since b is an inverse of a, i.e., $b \bullet a = e$
	=	c		for e is the identity element

The term b on the first line above and the c on the last are equal, since they are connected by a chain of equalities. In other words, there is only one inverse element of a. Similarly, to prove that the identity element of a group is unique, assume G is a group with two identity elements e and f. Then $e = e \bullet f = f$, hence e and f are equal.

Division

In groups, the existence of inverse elements implies that division is possible: given elements a and b of the group G, there is exactly one solution x in G to the equationx • a = b, namely b • a”1. In fact, we have (b • a”1) • a = b • (a”1 • a) = b • e = b.

Uniqueness results by multiplying the two sides of the equation x • a = b by a”1. The element b • a”1, often denoted b / a, is called the right quotient of b by a, or the result of the right division of b by a. Similarly there is exactly one solution y in G to the equation a • y = b, namely y = a”1 • b. This solution is the left quotient of b by a, and is sometimes denoted a \ b. In general b / a anda \ b may be different, but, if the group operation is commutative (that is, if the group is abelian), they are equal. In this case, the group operation is often denoted as an addition, and one talks of subtraction and difference instead of division and quotient. Abstract Algebra A consequence of this is that multiplication by a group element g is a bijection. Specifically, if g is an element of the group G, the function (mathematics) from G to itself that maps h"G to g • h is a bijection. This function is called the left translation by g. Similarly, the right translation by g is the bijection from G to itself, that maps h to h • g. If G is abelian, the left and the right translation by a group element are the same.

Basic Concepts

To understand groups beyond the level of mere symbolic manipulations as above, more structural concepts have to be employed.c[›] There is a conceptual principle underlying all of the following notions: to take advantage of the structure offered by groups (which sets, being "structureless", do not have), constructions related to groups have to be compatible with the group operation. This compatibility manifests itself in the following notions in various ways. For example, groups can be related to each other via functions called group homomorphisms. By the mentioned principle, they are required to respect the group structures in a precise sense. The structure of groups can also be understood by breaking them into pieces called subgroups and quotient groups. The principle of "preserving structures"—a recurring topic in mathematics throughout—is an instance of working in a category, in this case the category of groups.

Group Homomorphisms

Group homomorphismsg[›] are functions that preserve group structure. A function a: G '! H between two groups (G, •) and (H, ") is called a homomorphism if the equation a(g • k) = a(g) "a(k) holds for all elements g, k in G. In other words, the result is the same when performing the group operation after or before applying the map a. This requirement ensures that a(1G) = 1H, and also a(g)™1 = a(g™1) for all g in G. Thus a group homomorphism respects all the structure of G provided by the group axioms. Two groups G and H are called isomorphic if there exist group homomorphisms a: G '! H and b: H '! G, such that applying the two functions one after another in each of the two possible orders gives the identity functions of G and H. That is, a(b(h)) = h and b(a(g)) = g for any g in G and h in H. From an abstract point of view, isomorphic groups carry the same information. For example, proving that g • g = 1G for some element g of G is equivalent to proving that a(g) ™a(g) =

1H, because applying a to the first equality yields the second, and applying b to the second gives back the first.

Subgroups

Informally, a subgroup is a group H contained within a bigger one, G. Concretely, the identity element of G is contained in H, and whenever h1 and h2 are in H, then so are h1 • h2 and h1™1, so the elements of H, equipped with the group operation on G restricted to H, indeed form a group. In the example above, the identity and the rotations constitute a subgroup R = {id, r1, r2, r3}, highlighted in red in the group table above: any two rotations composed are still a rotation, and a rotation can be undone by (i.e., is inverse to) the complementary rotations 270° for 90°, 180° for 180°, and 90° for 270° (note that rotation in the opposite direction is not defined). The subgroup test is a necessary and sufficient condition for a nonempty subset H of a group G to be a subgroup: it is sufficient to check that g™1h™H for all elements g, h™H. Knowing the subgroups is important in understanding the group as a whole.d[›] Given any subset S of a group G, the subgroup generated by S consists of products of elements of S and their inverses. It is the smallest subgroup of G containing S. In the introductory example above, the subgroup generated by r2 and fv consists of these two elements, the identity element id and fh = fv • r2. Again, this is a subgroup, because combining any two of these four elements or their inverses (which are, in this particular case, these same elements) yields an element of this subgroup.

Cosets

In many situations it is desirable to consider two group elements the same if they differ by an element of a given subgroup. For example, in D4 above, once a reflection is performed, the square never gets back to the r2 configuration by just applying the rotation operations (and no further reflections), i.e., the rotation operations are irrelevant to the question whether a reflection has been performed. Cosets are used to formalize this insight: a subgroup H defines left and right cosets, which can be thought of as translations of H by arbitrary group elements g. In symbolic terms, the left and right cosets of H containing g are gH = {g • h : h™H} and Hg = {h • g : h™H}, respectively.

The left cosets of any subgroup H form a partition of G; that is, the union of all left cosets is equal to G and two left cosets are either equal or have an emptyintersection. The first case g1H = g2H happens precisely wheng1"1 • g2"H, i.e., if the two elements differ by an element of H. Similar considerations apply to the right cosets of H. The left and right cosets of H may or may not be equal. If they are, i.e., for all g in G, gH = Hg, then H is said to be a normal subgroup. In D4, the introductory symmetry group, the left cosets gR of the subgroup R consisting of the rotations are either equal to R, if g is an element of R itself, or otherwise equal to U = fcR = {fc, fv, fd, fh} (highlighted in green). The subgroup R is also normal, because fcR = U = Rfc and similarly for any element other than fc. (In fact, in the case of D4, observe that all such cosets are equal, such that fhR = fvR = fdR = fcR.)

Coset

G is the group (Z/8Z, +), the integers mod 8 under addition. The subgroup H contains only 0 and 4, and is isomorphic to (Z/2Z, +). There are four left cosets of H: H itself, 1 + H, 2 + H, and 3 + H (written using additive notation since this is the additive group). Together they partition the entire group G into equal-size, non-overlapping sets. The index [G : H] is 4. In mathematics, if G is a group, and H is a subgroup of G, and g is an element of G, then gH = is the left coset of H in G ith respect to g, and Hg = is the right coset of H in G with respect to g. Although derived from a subgroup, cosets are not usually themselves subgroups of G, only subsets. For example, it is not necessary for a coset to contain the identity element. A coset may be regarded as a left or a right coset of some subgroup in G. For example, the right coset of H with respect to g, Hg, is also the left coset of the conjugate subgroup g"1Hg with respect to g, g. In other words, a right coset of one subgroup equals a left coset of a conjugate subgroup. It is only when the subgroup (H in the above discussion) is a normal subgroup that they are the same. Hence it is not meaningful to speak of a coset as being left or right unless one first specifies the underlying subgroup. If (and only if) H is a normal subgroup of G, the cosets form a group called the quotient or factor group. If G is an abelian group, the left and right cosets will always be the same. The map gH ¦! (gH)"1 = Hg"1 defines a bijection between the left cosets and the right cosets of H, so the number of left cosets is equal to the number of right cosets. The common value is called the index of H in G. If the group operation is written additively, the notation used changes to g + H or H + g. Cosets are a basic tool in the study of groups; for example they play a central role in Lagrange's theorem.

Examples

C2Let G = ({"1,1}, ×) be the group formed by {"1,1} under multiplication, which is isomorphic to C2, and H the trivial subgroup ({1}, ×). Then {"1} = ("1)H = H("1) and {1} = 1H = H1 are the only cosets of H in G. Because its left and right cosets with respect to any element of G coincide, H is a normal subgroup of G.

Integers

Let G be the additive group of the integers, Z = ({..., “2, “1, 0, 1, 2,...}, +) and H the subgroup (mZ, +) = ({..., “2m, “m, 0, m, 2m,...}, +) where m is a positive integer. Then the cosets of H in G are the m sets mZ, mZ + 1,..., mZ + (m “ 1), where mZ + a = {..., “2m+a, “m+a, a, m+a, 2m+a,...}. There are no more than m cosets, because mZ + m = m(Z + 1) = mZ. The coset (mZ + a, +) is the congruence class of a modulo m.

Vectors

Another example of a coset comes from the theory of vector spaces. The elements (vectors) of a vector space form an abelian group under vector addition. It is not hard to show that subspaces of a vector space are subgroups of this group. For a vector space V, a subspace W, and a fixed vector a in V, the sets are called affine subspaces, and are cosets (both left and right, since the group is abelian). In terms of geometric vectors, these affine subspaces are all the “lines” or “planes” parallel to the subspace, which is a line or plane going through the origin.

DEFINITION USING EQUIVALENCE CLASSES

Some authors define the left cosets of H in G to be the equivalence classes under the equivalence relation on G given by x ~ y if and only if x”1y “ H. The relation can also be defined by x ~ y if and only if xh=y for some h in H. It can be shown that the relation given is, in fact, an equivalence relation and that the two definitions are equivalent. It follows that any two left cosets of H in G are either identical or disjoint. In other words, every element of G belongs to one and only one left coset and so the left cosets form a partition of G. Corresponding statements are true for right cosets.

Double Cosets

Given two subgroups, H and K of a group G, the double coset of H and K in G are sets of the form HgK = These are the left cosets of K and right cosets of H when H=1 and K=1 respectively.

Notation

Let G be a group with subgroups H and K.

1. denotes the set of left cosets of H in G.
2. denotes the set of right cosets of H in G.
3. denotes the set of double cosets of H and K in G.

General Properties

The identity is in precisely one left or right coset, namely H itself. Thus H is both a left and right coset of itself. A coset representative is a representative in the equivalence class sense. A set of representatives of all the cosets is called a transversal. There are other types of equivalence relations in a group, such as conjugacy, that form different classes which do not have the properties discussed here.

Index of a subgroup

All left cosets and all right cosets have the same order (number of elements, or cardinality in the case of an infinite H), equal to the order of H (because H is itself a coset). Furthermore, the number of left cosets is equal to the number of right cosets and is known as the index of H in G, written as [G : H]. Lagrange's theorem allows us to compute the index in the case where G and H are finite, as per the formula:

This equation also holds in the case where the groups are infinite, although the meaning may be less clear.

Cosets and Normality

If H is not normal in G, then its left cosets are different from its right cosets. That is, there is an a in G such that no element b satisfies aH = Hb. This means that the partition of G into the left cosets of H is a different partition than the partition of G into right cosets of H. (Some cosets may coincide. For example, if a is in the center of G, then aH = Ha.) On the other hand, the subgroup N is normal if and only if gN = Ng for all g in G. In this case, the set of all cosets form a group called the quotient group G / N with the operation " defined by (aN) " (bN) = abN. Since every right coset is a left coset, there is no need to distinguish "left cosets" from "right cosets".

Applications

- Cosets of Q in R are used in the construction of Vitali sets, a type of non-measurable set.
- Cosets are central in the definition of the transfer.
- Cosets are important in computational group theory. For example, Thistlethwaite's algorithm for solving Rubik's Cube relies heavily on cosets.
- Coset leaders are used in decoding received data in linear error-correcting codes

Quotient Groups

In some situations the set of cosets of a subgroup can be endowed with a group law, giving a quotient group or factor group. For this to be possible, the subgroup has to be normal. Given any normal subgroup N, the quotient group is defined by

G / N = {gN, gG}, “G modulo N”. This set inherits a group operation (sometimes called coset multiplication, or coset addition) from the original group G: (gN) • (hN) = (gh)N for all g and h in G. This definition is motivated by the idea (itself an instance of general structural considerations outlined above) that the map G ’! G / N that associates to any element g its coset gN be a group homomorphism, or by general abstract considerations called universal properties. The coset eN = N serves as the identity in this group, and the inverse of gN in the quotient group is (gN)”1 = (g”1)N.e[›]

Group table of the quotient group D4 / R

Group table of the quotient group D_4 / R		
•	**R**	**U**
R	*R*	*U*
U	*U*	*R*

The elements of the quotient group D4 / R are R itself, which represents the identity, and U = fvR. The group operation on the quotient is shown at the right. For example, U • U = fvR • fvR = (fv • fv)R = R. Both the subgroup R = {id, r1, r2, r3}, as well as the corresponding quotient are abelian, whereas D4 is not abelian. Building bigger groups by smaller ones, such as D4 from its subgroup

R and the quotient D4 / R is abstracted by a notion called semidirect product. Quotient groups and subgroups together form a way of describing every group by its presentation: any group is the quotient of the free group over the generators of the group, quotiented by the subgroup of relations. The dihedral group D4, for example, can be generated by two elements r and f (for example, r = r1, the right rotation and f = fv the vertical (or any other) reflection), which means that every symmetry of the square is a finite composition of these two symmetries or their inverses. Together with the relations

r4 = f 2 = (r • f)2 = 1, the group is completely described. A presentation of a group can also be used to construct the Cayley graph, a device used to graphically capture discrete groups.

Sub- and quotient groups are related in the following way: a subset H of G can be seen as an injective map H ’! G, i.e., any element of the target has at most one element that maps to it. The counterpart to injective maps aresurjective maps (every element of the target is mapped onto), such as the canonical map G '! G / N.y[›] Interpreting subgroup and quotients in light of these homomorphisms emphasizes the structural concept inherent to these definitions alluded to in the introduction. In general, homomorphisms are neither injective nor surjective. Kernel and image of group homomorphisms and the first isomorphism theorem address this phenomenon.

EXAMPLES AND APPLICATIONS

Examples and applications of groups abound. A starting point is the group Z of integers with addition as group operation, introduced above. If instead of addition multiplication is considered, one obtains multiplicative groups. These groups are predecessors of important constructions in abstract algebra.

Groups are also applied in many other mathematical areas. Mathematical objects are often examined by associating groups to them and studying the properties of the corresponding groups. For example, Henri Poincaré founded what is now called algebraic topology by introducing the fundamental group. By means of this connection, topological properties such as proximity and continuity translate into properties of groups.i[›] For example, elements of the fundamental group are represented by loops. The second image at the right shows some loops in a plane minus a point. The blue loop is considered null-homotopic (and thus irrelevant), because it can be continuously shrunk to a point. The presence of the hole prevents the orange loop from being shrunk to a point.

The fundamental group of the plane with a point deleted turns out to be infinite cyclic, generated by the orange loop (or any other loop winding once around the hole). This way, the fundamental group detects the hole. In more recent applications, the influence has also been reversed to motivate geometric constructions by a group-theoretical background.j[›] In a similar vein, geometric group theory employs geometric concepts, for example in the study of hyperbolic groups. Further branches crucially applying groups include algebraic geometry and number theory. In addition to the above theoretical applications, many practical applications of groups exist. Cryptography relies on the combination of the abstract group theory approach together with algorithmical knowledge obtained in computational group theory, in particular when implemented for finite groups. Applications of group theory are not restricted to mathematics; sciences such as physics, chemistry and computer science benefit from the concept.

Numbers

Many number systems, such as the integers and the rationals enjoy a naturally given group structure. In some cases, such as with the rationals, both addition and multiplication operations give rise to group structures. Such number systems are predecessors to more general algebraic structures known as rings and fields. Further abstract algebraic concepts such as modules, vector spaces and algebras also form groups.

Integers

The group of integers z under addition, denoted (Z, +), has been described above. The integers, with the operation of multiplication instead of addition, (Z, Å") do not form a group. The closure, associativity and identity axioms are satisfied, but inverses do not exist: for example, a = 2 is an integer, but the only solution to the equation $a \cdot b = 1$ in this case is b = 1/2, which is a rational number, but not an integer. Hence not every element of z has a (multiplicative) inverse.

Rationals

The desire for the existence of multiplicative inverses suggests considering fractions Fractions of integers (with b nonzero) are known as rational numbers.l[›] The set of all such irreducible fractions is commonly denoted Q.Θ There is still a minor obstacle for (Q, Å"), the rationals with multiplication, being a group: because the rational number 0 does not have a multiplicative inverse (i.e., there is no x such that $x \cdot 0 = 1$), (Θ, Å") is still not a group.

However, the set of all nonzero rational numbers $\Theta \therefore \{0\} = \{\theta \in \Theta \mid \theta \neq 0\}$ does form an abelian group under multiplication, generally denoted Q "Θ*. Associativity and identity element axioms follow from the properties of integers. The closure requirement still holds true after removing zero, because the product of two nonzero rationals is never zero. Finally, the inverse of a/b is b/a, therefore the axiom of the inverse element is satisfied.

The rational numbers (including 0) also form a group under addition. Intertwining addition and multiplication operations yields more complicated structures called rings and—if division is possible, such as in which occupy a central position in abstract algebra. Group theoretic arguments therefore underlie parts of the theory of those entities.

Modular arithmetic

In modular arithmetic, two integers are added and then the sum is divided by a positive integer called the modulus. The result of modular addition is the remainder of that division. For any modulus, n, the set of integers from 0 to n " 1 forms a group under modular addition: the inverse of any element a isn " a, and 0 is the identity element. This is familiar from the addition of hours on the face of a clock: if the hour hand is on 9 and is advanced 4 hours, it ends up on 1, as shown at the right. This is expressed by saying that 9 + 4 equals 1 "modulo 12" or, in symbols, 9 + 4 a" 1 modulo 12.

The group of integers modulo n is written or Z /nZ. For any prime numberp, there is also the multiplicative group of integers modulo p. Its elements are the integers 1 to p ™ 1. The group operation is multiplication modulo p. That is, the usual product is divided by p and the remainder of this division is the result of modular multiplication. For example, if p = 5, there are four group elements 1, 2, 3, 4. In this group, 4 · 4 = 1, because the usual product 16 is equivalent to 1, which divided by 5 yields a remainder of 1. for 5 divides 16 ™ 1 = 15, denoted 16 a™ 1 (mod 5). The primality of p ensures that the product of two integers neither of which is divisible by p is not divisible by peither, hence the indicated set of classes is closed under multiplication.o[›] The identity element is 1, as usual for a multiplicative group, and the associativity follows from the corresponding property of integers. Finally, the inverse element axiom requires that given an integer a not divisible by p, there exists an integer b such that a · b a" 1 (mod p), i.e., p divides the difference a · b " 1.

The inverse b can be found by using Bézout's identity and the fact that the greatest common divisorgcd(a, p) equals 1. In the case p = 5 above, the inverse of 4 is 4, and the inverse of 3 is 2, as 3 · 2 = 6 a" 1 (mod 5). Hence all group axioms are fulfilled. Actually, this example is similar to (Q " {0}, Å") : it consists of exactly those elements in Z/ πZ that have a multiplicative inverse. These groups are denoted Fp×. They are crucial to public-key cryptography.P[›]

Cyclic groups

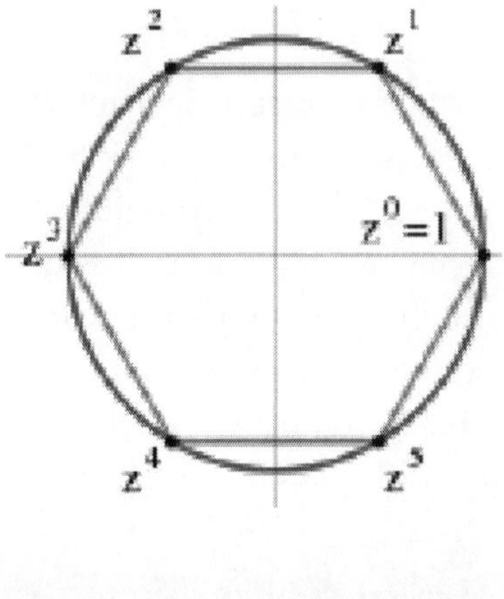

The 6th complex roots of unity form a cyclic group. z is a primitive element, but z2 is not, because the odd powers of z are not a power of z2. A cyclic group is a group all of whose elements are powers of a particular element a. In multiplicative notation, the elements of the group are:..., a”3, a”2, a”1, a0 = e, a, a2, a3,...,

wherea2 means a • a, and a”3 stands for a”1 • a”1 • a”1 = (a • a • a)”1 etc.h[›] Such an element a is called a generator or a primitive element of the group. In additive notation, the requirement for an element to be primitive is that each element of the group can be written as..., “a”a, “a, 0, a, a+a,...

In the groups Z/nZ introduced above, the element 1 is primitive, so these groups are cyclic. Indeed, each element is expressible as a sum all of whose terms are 1. Any cyclic group with n elements is isomorphic to this group. A second example for cyclic groups is the group of n-th complex roots of unity, given by complex numbersz satisfying zn = 1. These numbers can be visualized as the vertices on a regular n-gon, as shown in blue at the right for n = 6. The group operation is multiplication of complex numbers. In the picture, multiplying with z corresponds to a counter-clockwise rotation by 60°. Using some field theory, the group Fp× can be shown to be cyclic: for example, if p = 5, 3 is a generator since 31 = 3, 32 = 9 a” 4, 33 a” 2, and 34 a” 1.

Some cyclic groups have an infinite number of elements. In these groups, for every non-zero element a, all the powers of a are distinct; despite the name “cyclic group”, the powers of the elements do not cycle. An infinite cyclic group is isomorphic to (Z, +), the group of integers under addition introduced above. As these two prototypes are both abelian, so is any cyclic group. The study of finitely generated abelian groups is quite mature, including the fundamental theorem of finitely generated abelian groups; and reflecting this state of affairs, many group-related notions, such as center and commutator, describe the extent to which a given group is not abelian.

Symmetry groups

Symmetry groups are groups consisting of symmetries of given mathematical objects—be they of geometric nature, such as the introductory symmetry group of the square, or of algebraic nature, such as polynomial equations and their solutions. Conceptually, group theory can be thought of as the study of symmetry.Symmetries in mathematics greatly simplify the study of geometrical or analytical objects. A group is said to act on another mathematical object X if every group element performs some operation on X compatibly to the group law. In the rightmost example below, an element of order 7 of the (2,3,7) triangle group acts on the tiling by permuting the highlighted warped triangles (and the other ones, too). By a group action, the group pattern is connected to the structure of the object being acted on.

In chemical fields, such as crystallography, space groups and point groups describe molecular symmetries and crystal symmetries. These symmetries underlie the chemical and physical behavior of these systems, and group theory enables simplification of quantum mechanical analysis of these properties. For example, group theory is used to show that optical transitions between certain quantum levels cannot occur simply because of the symmetry of the states involved. Not only are groups useful to assess the implications of symmetries in molecules, but surprisingly they also predict that molecules sometimes can change symmetry. The Jahn-Teller effect is a distortion of a molecule of high symmetry when it adopts a particular ground state of lower symmetry from a set of possible ground states that are related to each other by the symmetry operations of the molecule.

Likewise, group theory helps predict the changes in physical properties that occur when a material undergoes a phase transition, for example, from a cubic to a tetrahedral crystalline form. An example is ferroelectric materials, where the change from a paraelectric to a ferroelectric state occurs at the Curie temperature and is related to a change from the high-symmetry paraelectric state to the lower symmetry ferroelectric state, accompanied by a so-called soft phonon mode, a vibrational lattice mode that goes to zero frequency at the transition. Such spontaneous symmetry breaking has found further application in elementary particle physics, where its occurrence is related to the appearance of Goldstone bosons.

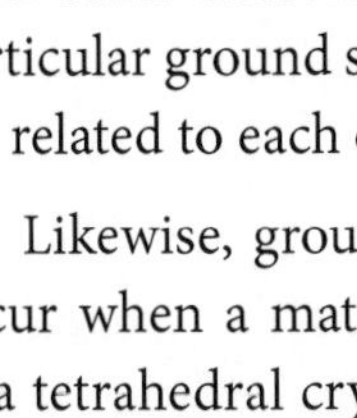	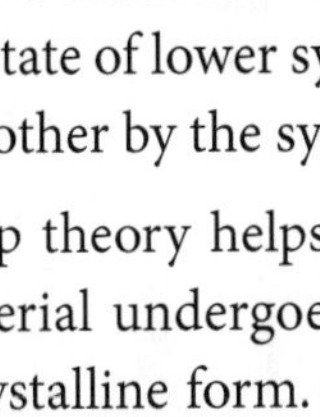	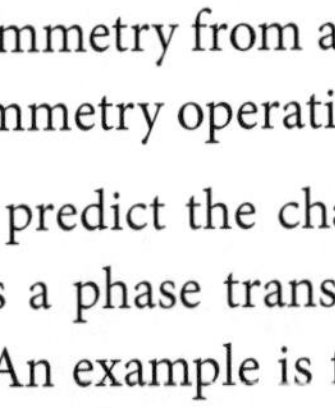	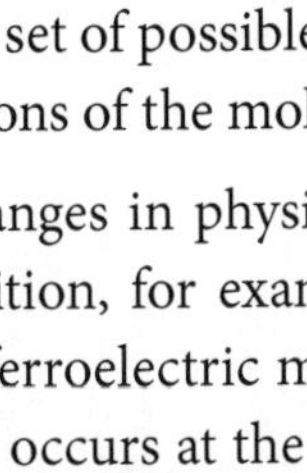	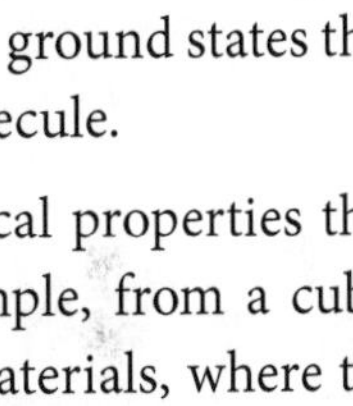
Buckminsterfullerene displays icosahedral symmetry, though the double bonds reduce this to pyritohedral symmetry.	Ammonia, NH_3. Its symmetry group is of order 6, generated by a 120° rotation and a reflection.	CubaneC_8H_8 features octahedral symmetry.	Hexaaquacopper(II) complex ion, $[Cu(OH_2)_6]^{2+}$. Compared to a perfectly symmetrical shape, the molecule is vertically dilated by about 22% (Jahn-Teller effect).	The (2,3,7) triangle group, a hyperbolic group, acts on this tiling of the hyperbolic plane.

Finite symmetry groups such as the Mathieu groups are used in coding theory, which is in turn applied in error correction of transmitted data, and in CD players. Another application is differential Galois theory, which characterizes functions having antiderivatives of a prescribed form, giving group-theoretic criteria for when solutions of certain differential equations are well-behaved.u[›] Geometric properties that remain stable under group actions are investigated in (geometric) invariant theory.

GENERAL LINEAR GROUP AND REPRESENTATION THEORY

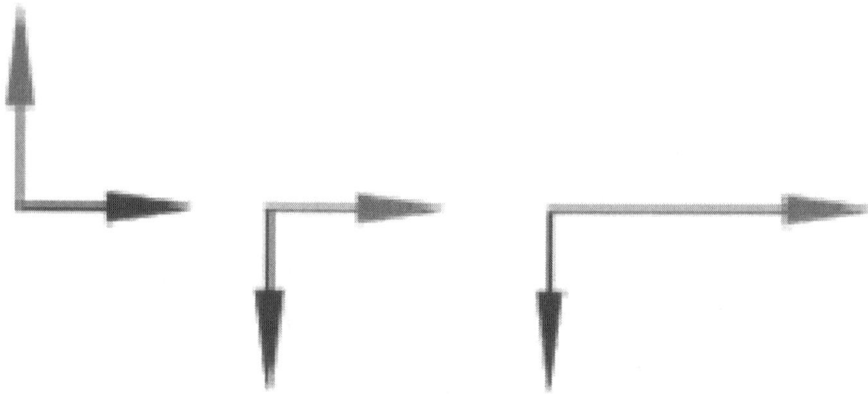

Two vectors (the left illustration) multiplied by matrices (the middle and right illustrations). The middle illustration represents a clockwise rotation by 90°, while the right-most one stretches the x-coordinate by factor 2. Matrix groups consist of matrices together with matrix multiplication. The general linear groupGL(n, R) consists of all invertiblen-by-n matrices with real entries. Its subgroups are referred to as matrix groups or linear groups. The dihedral group example mentioned above can be viewed as a (very small) matrix group. Another important matrix group is the special orthogonal groupSO(n). It describes all possible rotations in n dimensions. Via Euler angles, rotation matrices are used in computer graphics.

Representation theory is both an application of the group concept and important for a deeper understanding of groups. It studies the group by its group actions on other spaces. A broad class of group representations are linear representations, i.e., the group is acting on a vector space, such as the three-dimensional Euclidean spaceR3. A representation of G on an n-dimensional real vector space is simply a group homomorphism ñ: G '! GL(n, R) from the group to the general linear group. This way, the group operation, which may be abstractly given, translates to the multiplication of matrices making it accessible to explicit computations.Given a group action, this gives further means to study the object being acted on.On the other hand, it also yields information about the group. Group representations are an organizing principle in the theory of finite groups, Lie groups, algebraic groups and topological groups, especially (locally) compact groups.

Galois groups

Galois groups were developed to help solve polynomial equations by capturing their symmetry features. For example, the solutions of the quadratic equationax2 + bx + c = 0 are given by

$$x = \frac{-b \pm \sqrt{b^2 - 4ac}}{2a}$$

Exchanging "+" and """ in the expression, i.e., permuting the two solutions of the equation can be viewed as a (very simple) group operation. Similar formulae are known for cubic and quartic equations, but do not exist in general for degree 5 and higher. Abstract properties of Galois groups associated with polynomials (in particular their solvability) give a criterion for polynomials that have all their solutions expressible by radicals, i.e., solutions expressible using solely addition, multiplication, and roots similar to the formula above. The problem can be dealt with by shifting to field theory and considering the splitting field of a polynomial. Modern Galois theory generalizes the above type of Galois groups to field extensions and establishes—via the fundamental theorem of Galois theory—a precise relationship between fields and groups, underlining once again the ubiquity of groups in mathematics.

Finite Groups

A group is called finite if it has a finite number of elements. The number of elements is called the order of the group. An important class is the symmetric groups SN, the groups of permutations of N letters. For example, the symmetric group on 3 letters S3 is the group consisting of all possible orderings of the three letters ABC, i.e., contains the elements ABC, ACB, BAC, BCA, CAB, CBA, in total 6 (factorial of 3) elements. This class is fundamental insofar as any finite group can be expressed as a subgroup of a symmetric group SN for a suitable integer N, according to Cayley's theorem. Parallel to the group of symmetries of the square above, S3 can also be interpreted as the group of symmetries of an equilateral triangle. The order of an element a in a group G is the least positive integer n such that an = e, where an represents

$$\underbrace{a \ldots a}_{n\ factors},$$

i.e., application of the operation • to n copies of a. (If • represents multiplication, then an corresponds to the nth power of a.) In infinite groups, such an n may not exist, in which case the order of a is said to be infinity. The order of an element equals the order of the cyclic subgroup generated by this element. More sophisticated counting techniques, for example counting cosets, yield more precise statements about finite groups: Lagrange's Theorem states that for a finite group G the order of any finite subgroup Hdivides the order of G. The Sylow theorems give a partial converse. The dihedral group (discussed above) is a finite group of order 8. The order of r1 is 4, as is the order of the subgroup R it generates (see above). The order of the reflection elements fv etc. is 2. Both orders divide 8, as predicted by Lagrange's theorem. The groups Fp× above have order p " 1.

Classification of Finite Simple Groups

Mathematicians often strive for a complete classification (or list) of a mathematical notion. In the context of finite groups, this aim leads to difficult mathematics. According to Lagrange's theorem, finite groups of order p, a prime number, are necessarily cyclic (abelian) groups Zp. Groups of order p2 can also be shown to be abelian, a statement which does not generalize to order p3, as the non-abelian group D4 of order 8 = 23 above shows. Computer algebra systems can be used to list small groups, but there is no classification of all finite groups.q[›] An intermediate step is the classification of finite simple groups.r[›] A nontrivial group is called simple if its only normal subgroups are the trivial group and the group itself.s[›] The Jordan–Hölder theorem exhibits finite simple groups as the building blocks for all finite groups.

Listing all finite simple groups was a major achievement in contemporary group theory. 1998 Fields Medal winner Richard Borcherds succeeded in proving the monstrous moonshine conjectures, a surprising and deep relation between the largest finite simple sporadic group—the "monster group"—and certain modular functions, a piece of classical complex analysis, and string theory, a theory supposed to unify the description of many physical phenomena.

Groups with Additional Structure

Many groups are simultaneously groups and examples of other mathematical structures. In the language of category theory, they are group objects in a category, meaning that they are objects (that is, examples of another mathematical structure) which come with transformations (called morphisms) that mimic the group axioms. For example, every group (as defined above) is also a set, so a group is a group object in the category of sets.

Topological Groups

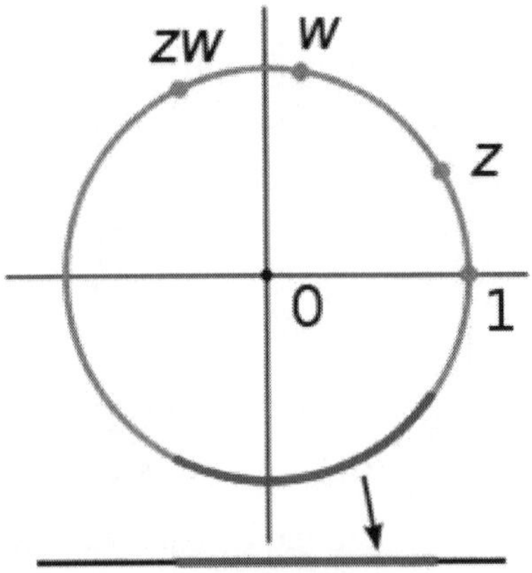

The unit circle in the complex plane under complex multiplication is a Lie group and, therefore, a topological group. It is topological since complex multiplication and division are continuous. It is a manifold and thus a Lie group, because every small piece, such as the red arc in the figure, looks like a part of the real line (shown at the bottom).

Some topological spaces may be endowed with a group law. In order for the group law and the topology to interweave well, the group operations must be continuous functions, that is, g • h, and g"1 must not vary wildly if g and h vary only little. Such groups are called topological groups, and they are the group objects in the category of topological spaces. The most basic examples are the realsR under addition, (R" {0}, ·), and similarly with any other topological field such as the complex numbers or p-adic numbers. All of these groups are locally compact, so they have Haar measures and can be studied via harmonic analysis. The former offer an abstract formalism of invariant integrals.Invariance means, in the case of real numbers for example:

$$\int f(x)dx = \int f(x+c)dx$$

for any constant c. Matrix groups over these fields fall under this regime, as do adele rings and adelic algebraic groups, which are basic to number theory. Galois groups of infinite field extensions such as the absolute Galois group can also be equipped with a topology, the so-called Krull topology, which in turn is central to generalize the above sketched connection of fields and groups to infinite field extensions. An advanced generalization of this idea, adapted to the needs of algebraic geometry, is the étale fundamental group.

Lie groups

Lie groups (in honor of Sophus Lie) are groups which also have a manifold structure, i.e., they are spaces looking locally like some Euclidean space of the appropriate dimension. Again, the additional structure, here the manifold structure, has to be compatible, i.e., the maps corresponding to multiplication and the inverse have to be smooth. A standard example is the general linear group introduced above: it is an open subset of the space of all n-by-n matrices, because it is given by the inequality

det (A) "' 0,

whereA denotes an n-by-n matrix.

Lie groups are of fundamental importance in modern physics: Noether's theorem links continuous symmetries to conserved quantities.Rotation, as well as translations in space and timeare basic symmetries of the laws of mechanics. They can, for instance, be used to construct simple models—imposing, say, axial symmetry on a situation will typically lead to significant simplification in the equations one needs to solve to provide a physical description.v[›] Another example are the Lorentz transformations, which relate measurements of time and velocity of two observers in motion relative to each other. They can be deduced in a purely group-theoretical way, by expressing the transformations as a rotational symmetry of Minkowski space. The latter serves—in the absence of significant gravitation—as a model of space time in special relativity. The full symmetry group of Minkowski space, i.e., including translations, is known as the Poincaré group. By the above, it plays a pivotal role in special relativity and, by implication, for quantum field theories.Symmetries that vary with location are central to the modern description of physical interactions with the help of gauge theory.

GENERALIZATIONS

Group-like structures					
	Totality[a]	Associativity	Identity	Invertibility	Commutativity
Semigroupoid	Unneeded	Required	Unneeded	Unneeded	Unneeded
Small Category	Unneeded	Required	Required	Unneeded	Unneeded
Groupoid	Unneeded	Required	Required	Required	Unneeded
Magma	Required	Unneeded	Unneeded	Unneeded	Unneeded
Quasigroup	Required	Unneeded	Unneeded	Required	Unneeded
Loop	Required	Unneeded	Required	Required	Unneeded
Semigroup	Required	Required	Unneeded	Unneeded	Unneeded
Inverse Semigroup	Required	Required	Unneeded	Required	Unneeded
Monoid	Required	Required	Required	Unneeded	Unneeded
Group	Required	Required	Required	Required	Unneeded
Abelian group	Required	Required	Required	Required	Required

^aClosure, which is used in many sources, is an equivalent axiom to totality, though defined differently.

In abstract algebra, more general structures are defined by relaxing some of the axioms defining a group. For example, if the requirement that every element has an inverse is eliminated, the resulting algebraic structure is called a monoid. The natural numbersN (including 0) under addition form a monoid, as do the nonzero integers under multiplication (Z" {0}, ·), see above. There is a general method to formally add inverses to elements to any (abelian) monoid, much the same way as (Q" {0}, ·) is derived from (Z" {0}, ·), known as the Grothendieck group. Groupoids are similar to groups except that the composition a • b need not be defined for all a andb. They arise in the study of more complicated forms of symmetry, often in topological and analytical structures, such as the fundamental groupoid or stacks. Finally, it is possible to generalize any of these concepts by replacing the binary operation with an arbitrary n-ary one (i.e., an operation taking n arguments). With the proper generalization of the group axioms this gives rise to an n-ary group. The table gives a list of several structures generalizing groups.

Connection of groups and symmetry

Given a structured object X of any sort, a symmetry is a mapping of the object onto itself which preserves the structure. This occurs in many cases, for example

If X is a set with no additional structure, a symmetry is a bijective map from the set to itself, giving rise to permutation groups.

If the object X is a set of points in the plane with its metric structure or any other metric space, a symmetry is a bijection of the set to itself which preserves the distance between each pair of points (an isometry). The corresponding group is called isometry group of X.

If instead angles are preserved, one speaks of conformal maps. Conformal maps give rise to Kleinian groups, for example.

Symmetries are not restricted to geometrical objects, but include algebraic objects as well. For instance, the equation x 2 - 3 = 0 {x^{2}-3=0} x^{2}-3=0 has the two solutions 3 {{\sqrt {3}}} {\sqrt {3}} and - 3 {-{\sqrt {3}}} -{\sqrt {3}}. In this case, the group that exchanges the two roots is the Galois group belonging to the equation. Every polynomial equation in one variable has a Galois group, that is a certain permutation group on its roots.

The axioms of a group formalize the essential aspects of symmetry. Symmetries form a group: they are closed because if you take a symmetry of an object, and then apply another symmetry, the result will still be a symmetry. The identity keeping the object fixed is always a symmetry of an object. Existence of inverses is guaranteed by undoing the symmetry and the associativity comes from the fact that symmetries are functions on a space, and composition of functions is associative.

Frucht's theorem says that every group is the symmetry group of some graph. So every abstract group is actually the symmetries of some explicit object.

The saying of "preserving the structure" of an object can be made precise by working in a category. Maps preserving the structure are then the morphisms, and the symmetry group is the automorphism group of the object in question.

Applications of group theory

Applications of group theory abound. Almost all structures in abstract algebra are special cases of groups. Rings, for example, can be viewed as abelian groups (corresponding to addition) together with a second operation (corresponding to multiplication). Therefore, group theoretic arguments underlie large parts of the theory of those entities.

Galois theory

Galois theory uses groups to describe the symmetries of the roots of a polynomial (or more precisely the automorphisms of the algebras generated by these roots). The fundamental theorem of Galois theory provides a link between algebraic field extensions and group theory. It gives an effective criterion for the solvability of polynomial equations in terms of the solvability of the corresponding Galois group. For example, S5, the symmetric group in 5 elements, is not solvable which implies that the general quintic equation cannot be solved by radicals in the way equations of lower degree can. The theory, being one of the historical roots of group theory, is still fruitfully applied to yield new results in areas such as class field theory.

Algebraic topology

Algebraic topology is another domain which prominently associates groups to the objects the theory is interested in. There, groups are used to describe certain invariants of topological spaces. They are called "invariants" because they are defined in such a way that they do not change if the space is subjected to some deformation. For example, the fundamental group "counts" how many paths in the space are essentially different. The Poincaré conjecture, proved in 2002/2003 by Grigori Perelman, is a prominent application of this idea. The influence is not unidirectional, though. For example, algebraic topology makes use of Eilenberg–MacLane spaces which are spaces with prescribed homotopy groups. Similarly algebraic K-theory relies in a way on classifying spaces of groups. Finally, the name of the torsion subgroup of an infinite group shows the legacy of topology in group theory.

Algebraic geometry

Algebraic geometry likewise uses group theory in many ways. Abelian varieties have been introduced above. The presence of the group operation yields additional information which makes these varieties particularly accessible. They also often serve as a test for new conjectures. (For example the Hodge conjecture (in certain cases).) The one-dimensional case, namely elliptic curves is studied in particular detail. They are both theoretically and practically intriguing. In another direction, toric varieties are algebraic varieties acted on by a torus. Toroidal embeddings have recently led to advances in algebraic geometry, in particular resolution of singularities.

Harmonic analysis

Analysis on Lie groups and certain other groups is called harmonic analysis. Haar measures, that is, integrals invariant under the translation in a Lie group, are used for pattern recognition and other image processing techniques.

Combinatorics

In combinatorics, the notion of permutation group and the concept of group action are often used to simplify the counting of a set of objects; see in particular Burnside's lemma.

Music

The presence of the 12-periodicity in the circle of fifths yields applications of elementary group theory in musical set theory. Transformational theory models musical transformations as elements of a mathematical group.

Physics

In physics, groups are important because they describe the symmetries which the laws of physics seem to obey. According to Noether's theorem, every continuous symmetry of a physical system corresponds to a conservation law of the system. Physicists are very interested in group representations, especially of Lie groups, since these representations often point the way to the "possible" physical theories. Examples of the use of groups in physics include the Standard Model, gauge theory, the Lorentz group, and the Poincaré group.

Group theory can be used to resolve the incompleteness of the statistical interpretations of mechanics developed by Willard Gibbs, relating to the summing of an infinite number of probabilities to yield a meaningful solution.

Chemistry and materials science

In chemistry and materials science, point groups are used to classify regular polyhedra, and the symmetries of molecules, and space groups to classify crystal structures. The assigned groups can then be used to determine physical properties (such as chemical polarity and chirality), spectroscopic properties (particularly useful for Raman spectroscopy, infrared spectroscopy, circular dichroism spectroscopy, magnetic circular dichroism spectroscopy, UV/Vis spectroscopy, and fluorescence spectroscopy), and to construct molecular orbitals.

Molecular symmetry is responsible for many physical and spectroscopic properties of compounds and provides relevant information about how chemical reactions occur. In order to assign a point group for any given molecule, it is necessary to find the set of symmetry operations present on it. The symmetry operation is an action, such as a rotation around an axis or a reflection through a mirror plane. In other words, it is an operation that moves the molecule such that it is indistinguishable from the original configuration. In group theory, the rotation axes and mirror planes are called "symmetry elements". These elements can be a point, line or plane with respect to which the symmetry operation is carried out. The symmetry operations of a molecule determine the specific point group for this molecule.

In chemistry, there are five important symmetry operations. They are identity operation (E), rotation operation or proper rotation (Cn), reflection operation (s), inversion (i) and rotation reflection operation or improper rotation (Sn). The identity operation (E) consists of leaving the molecule as it is. This is equivalent to any number of full rotations around any axis. This is a symmetry of all molecules, whereas the symmetry group of a chiral molecule consists of only the identity operation. An identity operation is a characteristic of every molecule even if it has no symmetry. Rotation around an axis (Cn) consists of rotating the molecule around a specific axis by a specific angle. It is rotation through the angle 360°/n, where n is an integer, about a rotation axis. For example, if a water molecule rotates 180° around the axis that passes through the oxygen atom and between the hydrogen atoms, it is in the same configuration as it started. In this case, n = 2, since applying it twice produces the identity operation. In molecules with more than one rotation axis, the Cn axis having the largest value of n is the highest order rotation axis or principal axis. For example in boron trifluoride (BF3), the highest order of rotation axis is C3, so the principal axis of rotation is C3.

In the reflection operation (s) many molecules have mirror planes, although they may not be obvious. The reflection operation exchanges left and right, as if each point had moved perpendicularly through the plane to a position exactly as far from the plane as when it started. When the plane is perpendicular to the principal axis of rotation, it is called sh (horizontal). Other planes, which contain the principal axis of rotation, are labeled vertical (sv) or dihedral (sd).

Inversion (i) is a more complex operation. Each point moves through the center of the molecule to a position opposite the original position and as far from the central point as where it started. Many molecules that seem at first glance to have an inversion center do not; for example, methane and other tetrahedral molecules lack inversion symmetry. To see this, hold a methane model with two hydrogen atoms in the vertical plane on the right and two hydrogen atoms in the horizontal plane on the left. Inversion results in two hydrogen atoms in the horizontal plane on the right and two hydrogen atoms in the vertical plane on the left. Inversion is therefore not a symmetry operation of methane, because the orientation of the molecule following the inversion operation differs from the original orientation. And the last operation is improper rotation or rotation reflection operation (Sn) requires rotation of 360°/n, followed by reflection through a plane perpendicular to the axis of rotation.

Cryptography

Very large groups of prime order constructed in elliptic curve cryptography serve for public-key cryptography. Cryptographical methods of this kind benefit from the flexibility of the geometric objects, hence their group structures, together with the complicated structure of these groups, which make the discrete logarithm very hard to calculate. One of the earliest encryption protocols, Caesar's cipher, may also be interpreted as a (very easy) group operation. Most cryptographic schemes use groups in some way. In particular Diffie–Hellman key exchange uses finite cyclic groups. So the term group-based cryptography refers mostly to cryptographic protocols that use infinite nonabelian groups such as a braid group.

7

GROUP THEORY RESEARCH TOPICS

In addition to developments in number theory and algebraic geometry, modern algebra has important applications to symmetry by means of group theory. The word group often refers to a group of operations, possibly preserving the symmetry of some object or an arrangement of like objects. In the latter case the operations are called permutations, and one talks of a group of permutations, or simply a permutation group. If a and ß are operations, their composite (a followed by ß) is usually written aß, and their composite in the opposite order (ß followed by a) is written ßa. In general, aß and ßa are not equal.

A group can also be defined axiomatically as a set with multiplication that satisfies the axioms for closure, associativity, identity element, and inverses (axioms 1, 6, 9, and 10). In the special case where aß and ßa are equal for all a and ß, the group is called commutative, or Abelian; for such Abelian groups, operations are sometimes written a + ß instead of aß, using addition in place of multiplication.

The first application of group theory was by the French mathematician Évariste Galois (1811–32) to settle an old problem concerning algebraic equations. The question was to decide whether a given equation could be solved using radicals (meaning square roots, cube roots, and so on, together with the usual operations of arithmetic). By using the group of all "admissible" permutations of the solutions, now known as the Galois group of the equation, Galois showed whether or not the solutions could be expressed in terms of radicals. His was the first important use of groups, and he was the first to use the term in its modern technical sense. It was many years before his work was fully understood, in part because of its highly innovative character and in part because he was not around to explain his ideas—at the age of 20 he was mortally wounded in a duel. The subject is now known as Galois theory.

Group theory developed first in France and then in other European countries during the second half of the 19th century. One early and essential idea was that many groups, and in particular all finite groups, could be decomposed into simpler groups in an essentially unique way. These simpler groups could not be decomposed further, and so they were called "simple," although their lack of further decomposition often makes them rather complex. This is rather like decomposing a whole number into a product of prime numbers, or a molecule into atoms.

In 1963 a landmark paper by the American mathematicians Walter Feit and John Thompson showed that if a finite simple group is not merely the group of rotations of a regular polygon, then it must have an even number of elements. This result was immensely important because it showed that such groups had to have some elements x such that x2 = 1. Using such elements enabled mathematicians to get a handle on the structure of the whole group. The paper led to an ambitious program for finding all finite simple groups that was completed in the early 1980s. It involved the discovery of several new simple groups, one of which, the "Monster," cannot operate in fewer than 196,883 dimensions. The Monster still stands as a challenge today because of its intriguing connections with other parts of mathematics.

In mathematics and abstract algebra, group theory studies the algebraic structures known as groups. The concept of a group is central to abstract algebra: other well-known algebraic structures, such as rings, fields, and vector spaces, can all be seen as groups endowed with additional operations and axioms. Groups recur throughout mathematics, and the methods of group theory have influenced many parts of algebra.

Linear algebraic groups and Lie groups are two branches of group theory that have experienced advances and have become subject areas in their own right. Various physical systems, such as crystals and the hydrogen atom, may be modelled by symmetry groups. Thus group theory and the closely related representation theory have many important applications in physics, chemistry, and materials science. Group theory is also central to public key cryptography.

CATEGORYTHEORY

Schematic representation of a category with objects X, Y, Z and morphisms f, g, g"'f. (The category's three identity morphisms 1X, 1Y and 1Z, if explicitly represented, would appear as three arrows, from the letters X, Y, and Z to themselves, respectively.) Category theory formalizes mathematical structure and its concepts in terms of a labeleddirected graph called a category, whose nodes are called objects, and whose labelled directed edges are called arrows (or morphisms). A category has two basic properties: the ability to compose the arrows associatively, and the existence of an identity arrow for each object.

The language of category theory has been used to formalize concepts of other high-level abstractions such as sets, rings, and groups. Informally, category theory is a general theory of functions. Several terms used in category theory, including the term "morphism", are used differently from their uses in the rest of mathematics. In category theory, morphisms obey conditions specific to category theory itself. Samuel Eilenberg and Saunders Mac Lane introduced the concepts of categories, functors, and natural transformations in 1942–45 in their study of algebraic topology, with the goal of understanding the processes that preserve mathematical structure. Category theory has practical applications in programming language theory, for example the usage of monads in functional programming. It may also be used as an axiomatic foundation for mathematics, as an alternative to set theory and other proposed foundations.

Basic concepts

Categories represent abstractions of other mathematical concepts. Many areas of mathematics can be formalised by category theory as categories. Hence category theory uses abstraction to make it possible to state and prove many intricate and subtle mathematical results in these fields in a much simpler way. A basic example of a category is the category of sets, where the objects are sets and the arrows are functions from one set to another. However, the objects of a category need not be sets, and the arrows need not be functions.

Any way of formalising a mathematical concept such that it meets the basic conditions on the behaviour of objects and arrows is a valid category—and all the results of category theory apply to it. The "arrows" of category theory are often said to represent a process connecting two objects, or in many cases a "structure-preserving" transformation connecting two objects. There are, however, many applications where much more abstract concepts are represented by objects and morphisms. The most important property of the arrows is that they can be "composed", in other words, arranged in a sequence to form a new arrow.

Applications of categories

Categories now appear in many branches of mathematics, some areas of theoretical computer science where they can correspond to types or to database schemas, and mathematical physics where they can be used to describe vector spaces. Probably the first application of category theory outside pure mathematics was the "metabolism-repair" model of autonomous living organisms by Robert Rosen.

Utility

Categories, objects, and morphisms

The study of categories is an attempt to axiomatically capture what is commonly found in various classes of related mathematical structures by relating them to the structure-preserving functions between them. A systematic study of category theory then allows us to prove general results about any of these types of mathematical structures from the axioms of a category. Consider the following example. The classGrp of groups consists of all objects having a "group structure". One can proceed to provetheorems about groups by making logical deductions from the set of axioms defining groups. For example, it is immediately proven from the axioms that the identity element of a group is unique.

Instead of focusing merely on the individual objects (e.g., groups) possessing a given structure, category theory emphasizes the morphisms – the structure-preserving mappings – between these objects; by studying these morphisms, one is able to learn more about the structure of the objects. In the case of groups, the morphisms are the group homomorphisms. A group homomorphism between two groups "preserves the group structure" in a precise sense; informally it is a "process" taking one group to another, in a way that carries along information about the structure of the first group into the second group. The study of group homomorphisms then provides a tool for studying general properties of groups and consequences of the group axioms. A similar type of investigation occurs in many mathematical theories, such as the study of continuous maps (morphisms) between topological spaces in topology (the associated category is called Top), and the study of smooth functions (morphisms) in manifold theory. Not all categories arise as "structure preserving (set) functions", however; the standard example is the category of homotopies between pointed topological spaces. If one axiomatizes relations instead of functions, one obtains the theory of allegories.

Functors

A category is itself a type of mathematical structure, so we can look for "processes" which preserve this structure in some sense; such a process is called a functor. Diagram chasing is a visual method of arguing with abstract "arrows" joined in diagrams. Functors are represented by arrows between categories, subject to specific defining commutativity conditions. Functors can define (construct) categorical diagrams and sequences (viz. Mitchell, 1965). A functor associates to every object of one category an object of another category, and to every morphism in the first category a morphism in the second. As a result, this defines a category of categories and functors – the objects are categories, and the morphisms (between categories) are functors. Studying categories and functors is not just studying a class of mathematical structures and the morphisms between them but rather the relationships between various classes of mathematical structures. This fundamental idea first surfaced in algebraic topology. Difficult topological questions can be translated into algebraic questions which are often easier to solve. Basic constructions, such as the fundamental group or the fundamental groupoid of a topological space, can be expressed as functors to the category of groupoids in this way, and the concept is pervasive in algebra and its applications.

Natural transformations

Abstracting yet again, some diagrammatic and/or sequential constructions are often "naturally related" – a vague notion, at first sight. This leads to the clarifying concept of natural transformation, a way to "map" one functor to another. Many important constructions in mathematics can be studied in this context. "Naturality" is a principle, like general covariance in physics, that cuts deeper than is initially apparent. An arrow between two functors is a natural transformation when it is subject to certain naturality or commutativity conditions. Functors and natural transformations ('naturality') are the key concepts in category theory.

Categories

A categoryC consists of the following three mathematical entities:

- A class ob(C), whose elements are called objects;
- A class hom(C), whose elements are called morphisms or maps or arrows. Each morphism f has a source object a and target object b.

The expression f :a → b, would be verbally stated as "f is a morphism from a to b". The expression hom(a, b) – alternatively expressed as homC(a, b), mor(a, b), or C(a, b) –

- Denotes the hom-class of all morphisms from a to b.
- A binary operation"", called composition of morphisms, such that for any three objects a, b, and c, we have : hom(b, c) × hom(a, b) hom(a, c). The composition of f : a → b and g : b → c is written as gf or gf,[a] governed by two axioms:
    - o Associativity: If f : a → b,g : b → candh : c → dthenh o (g o f) = (h o g)o f, and

- o Identity: For every object x, there exists a morphism $1_x : x \rightarrow x$ called the identity morphism for x, such that for every morphism $f : a \rightarrow b$, we have $1_b f = f = f\, 1_a$.

From the axioms, it can be proved that there is exactly one identity morphism for every object. Some authors deviate from the definition just given by identifying each object with its identity morphism.

Morphisms

Relations among morphisms (such as fg = h) are often depicted using commutative diagrams, with "points" (corners) representing objects and "arrows" representing morphisms. Morphisms can have any of the following properties. A morphism $f{:}a \rightarrow b$ is a:

- Monomorphism (or monic) if $f \circ g_1 = f \circ g_2$ implies $g_1 = g_2$ for all morphisms $g_1, g_2 : x \rightarrow a$.
- Epimorphism (or epic) if $g_1 f = g_2 f$ implies $g_1 = g_2$ for all morphisms $g_1, g_2 : b \rightarrow x$.
- Bimorphism if f is both epic and monic.
- Isomorphism if there exists a morphism $g : b \rightarrow a$ such that fg = 1b and gf $= 1_a$.[b]
- Endomorphism if a = b. end(a) denotes the class of endomorphisms of a.
- Automorphism if f is both an endomorphism and an isomorphism. aut(a) denotes the class of automorphisms of a.
- Retraction if a right inverse of f exists, i.e. if there exists a morphism $g : b \rightarrow a$ with $fg = 1_b$.
- Section if a left inverse of f exists, i.e. if there exists a morphism $g : b \rightarrow a$ with $gf = 1_a$.

Every retraction is an epimorphism, and every section is a monomorphism. Furthermore, the following three statements are equivalent:

- f is a monomorphism and a retraction;
- f is an epimorphism and a section;
- f is an isomorphism.

Functors

Functors are structure-preserving maps between categories. They can be thought of as morphisms in the category of all (small) categories. A (covariant) functor F from a category C to a category D, written $F : C \rightarrow D$, consists of:

- For each object x in C, an object F(x) in D; and
- For each morphism $f : x \to y$ in C, a morphism $F(f) : F(x) \to F(y)$,

such that the following two properties hold:

- For every object x in C, $F(1_x) = 1_{F(x)}$;
- For all morphisms $f : x \to y$ and $g : y\ z$, $F(gf) = F(g)\ F(f)$.

A contravariant functor F: C → D is like a covariant functor, except that it "turns morphisms around" ("reverses all the arrows"). More specifically, every morphism $f : x \to y$ in C must be assigned to a morphism $F(f) : F(y) \to F(x)$ in D. In other words, a contravariant functor acts as a covariant functor from the opposite categoryCop to D.

Natural transformations

A natural transformation is a relation between two functors. Functors often describe "natural constructions" and natural transformations then describe "natural homomorphisms" between two such constructions. Sometimes two quite different constructions yield "the same" result; this is expressed by a natural isomorphism between the two functors. If F and G are (covariant) functors between the categories C and D, then a natural transformation ç from F to G associates to every object X in C a morphism çX :F(X) → G(X) in D such that for every morphism f : X → Y in C, we have çYF(f) = G(f) çX; this means that the following diagram is commutative:

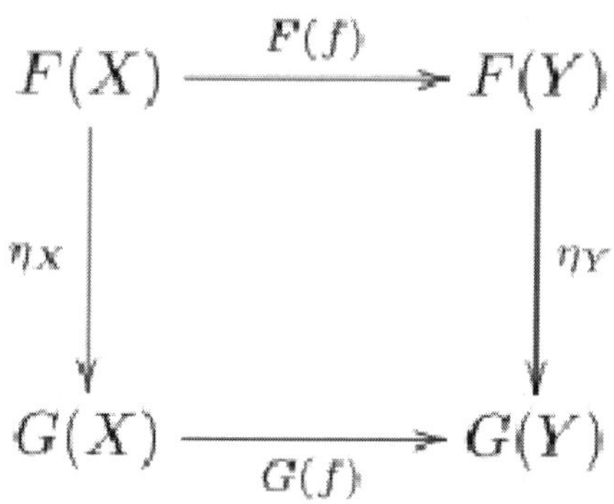

The two functors F and G are called naturally isomorphic if there exists a natural transformation from F to G such that çX is an isomorphism for every object X in C.

OTHER CONCEPTS

Universal constructions, limits, and colimits

Using the language of category theory, many areas of mathematical study can be categorized. Categories include sets, groups and topologies. Each category is distinguished by properties that all its objects have in common, such as the empty set or the product of two topologies, yet in the definition of a category, objects are considered atomic, i.e., we do not know whether an object A is a set, a topology, or any other abstract concept.

Hence, the challenge is to define special objects without referring to the internal structure of those objects. To define the empty set without referring to elements, or the product topology without referring to open sets, one can characterize these objects in terms of their relations to other objects, as given by the morphisms of the respective categories. Thus, the task is to find universal properties that uniquely determine the objects of interest. Numerous important constructions can be described in a purely categorical way if the category limit can be developed and dualized to yield the notion of a colimit.

Equivalent categories

It is a natural question to ask: under which conditions can two categories be considered essentially the same, in the sense that theorems about one category can readily be transformed into theorems about the other category? The major tool one employs to describe such a situation is called equivalence of categories, which is given by appropriate functors between two categories. Categorical equivalence has found numerous applications in mathematics.

Further concepts and results

The definitions of categories and functors provide only the very basics of categorical algebra; additional important topics are listed below. Although there are strong interrelations between all of these topics, the given order can be considered as a guideline for further reading.

- The functor category DC has as objects the functors from C to D and as morphisms the natural transformations of such functors. The Yoneda lemma is one of the most famous basic results of category theory; it describes representable functors in functor categories.
- Duality: Every statement, theorem, or definition in category theory has a dual which is essentially obtained by "reversing all the arrows". If one statement is true in a category C then its dual is true in the dual category Cop. This duality, which is transparent at the level of category theory, is often obscured in applications and can lead to surprising relationships.
- Adjoint functors: A functor can be left (or right) adjoint to another functor that maps in the opposite direction. Such a pair of adjoint functors typically

arises from a construction defined by a universal property; this can be seen as a more abstract and powerful view on universal properties.

Higher-dimensional categories

Many of the above concepts, especially equivalence of categories, adjoint functor pairs, and functor categories, can be situated into the context of higher-dimensional categories. Briefly, if we consider a morphism between two objects as a "process taking us from one object to another", then higher-dimensional categories allow us to profitably generalize this by considering "higher-dimensional processes".

For example, a (strict) 2-category is a category together with "morphisms between morphisms", i.e., processes which allow us to transform one morphism into another. We can then "compose" these "bimorphisms" both horizontally and vertically, and we require a 2-dimensional "exchange law" to hold, relating the two composition laws. In this context, the standard example is Cat, the 2-category of all (small) categories, and in this example, bimorphisms of morphisms are simply natural transformations of morphisms in the usual sense. Another basic example is to consider a 2-category with a single object; these are essentially monoidal categories. Bicategories are a weaker notion of 2-dimensional categories in which the composition of morphisms is not strictly associative, but only associative "up to" an isomorphism. This process can be extended for all natural numbers n, and these are called n-categories.

There is even a notion of ù-category corresponding to the ordinal number ù. Higher-dimensional categories are part of the broader mathematical field of higher-dimensional algebra, a concept introduced by Ronald Brown. For a conversational introduction to these ideas, see John Baez, 'A Tale of n-categories' (1996).

Historical notes

In 1942–45, Samuel Eilenberg and Saunders Mac Lane introduced categories, functors, and natural transformations as part of their work in topology, especially algebraic topology. Their work was an important part of the transition from intuitive and geometric homology to axiomatic homology theory. Eilenberg and Mac Lane later wrote that their goal was to understand natural transformations. That required defining functors, which required categories. Stanislaw Ulam, and some writing on his behalf, have claimed that related ideas were current in the late 1930s in Poland. Eilenberg was Polish, and studied mathematics in Poland in the 1930s.

Category theory is also, in some sense, a continuation of the work of Emmy Noether (one of Mac Lane's teachers) in formalizing abstract processes; Noether realized that understanding a type of mathematical structure requires understanding the processes that preserve that structure. To achieve this understanding, Eilenberg and Mac Lane proposed an axiomatic formalization of the relation between structures and the processes that preserve them.

The subsequent development of category theory was powered first by the computational needs of homological algebra, and later by the axiomatic needs of algebraic geometry. General category theory, an extension of universal algebra having many new features allowing for semantic flexibility and higher-order logic, came later; it is now applied throughout mathematics.

Certain categories called topoi (singular topos) can even serve as an alternative to axiomatic set theory as a foundation of mathematics. A topos can also be considered as a specific type of category with two additional topos axioms. These foundational applications of category theory have been worked out in fair detail as a basis for, and justification of, constructive mathematics. Topos theory is a form of abstract sheaf theory, with geometric origins, and leads to ideas such as pointless topology.

Categorical logic is now a well-defined field based on type theory for intuitionistic logics, with applications in functional programming and domain theory, where a cartesian closed category is taken as a non-syntactic description of a lambda calculus. At the very least, category theoretic language clarifies what exactly these related areas have in common (in some abstract sense). Category theory has been applied in other fields as well. For example, John Baez has shown a link between Feynman diagrams in Physics and monoidal categories. Another application of category theory, more specifically: topos theory, has been made in mathematical music theory, see for example the book The Topos of Music, Geometric Logic of Concepts, Theory, and Performance by Guerino Mazzola. More recent efforts to introduce undergraduates to categories as a foundation for mathematics include those of William Lawvere and Rosebrugh (2003) and Lawvere and Stephen Schanuel (1997) and Mirroslav Yotov (2012).

8

THE DEVELOPMENT OF GALOIS THEORY

Modern algebra may be described as the study of the structure of sets with certain properties and their interaction with one another via specific operations. However, merely 200 years ago, algebraists were primarily, if not exclusively, concerned with solving equations. The evolution of algebra from what is sometimes called "school algebra" today to the study of abstract structures largely coincides with the development of Galois Theory in the 19th and early 20th century. The concept of a group is generally credited to the French mathematician Évariste Galois, and while the idea of a field was developed by German mathematicians such as Kronecker and Dedekind, Galois Theory is what connects these two central concepts in algebra, the group and the field.

In 1971, B. Kiernan published an article on The Development of Galois Theory from Lagrange to Artin. While this article is a very well researched and comprehensive study, it runs over 114 pages and so a summary of it would be useful to have as a first point of reference. This project attempts to give such a summary and will thus contain very few original points, but instead try to summarise Kiernan's main points.

To put Galois' work into perspective, consider the historical background of the mathematical community to which Galois presented his ideas. During his lifetime, algebraists were interested in the solvability of equations of degree higher than four. The formula for the solution of the quadratic equation had essentially been known to the Babylonians and the general equations of degree three and four were solved during the Renaissance in Italy. Yet no one managed to find a general solution of the quintic and in the early 1800s, Ruffini and Abel proved that this was impossible.

However, this general result has no consequences for the algebraic solvability of any specific equation of degree five, and algebraists were seeking criteria which would determine whether a given quintic was solvable by radicals.

It was in this context that Galois published his work. Today, we see Galois Theory as the study of the structure of fields and their automorphism groups and although Galois used different terminology, this was essentially how he conceived his theory. As one possible application, he described how his theory could be used to determine necessary and sufficient conditions for the solvability of an equation by radicals.

Although, Galois pointed out that he was presenting "the general principles and just one application" of his theory, it is not surprising, given the historical context, that the mathematicians of his time concentrated on this application and largely ignored the general theory. They were trying to extract from Galois' work a procedure for finding the solutions of any given special equation, although Galois himself had never intended for his theory to yield such a procedure. It was only much later in the 19th century that mathematicians became interested in Galois Theory for its own sake, beyond its application to the solvability of equations. This development of Galois Theory will now be looked at in some greater detail.

APPLICATION TO CLASSICAL PROBLEMS

The birth and development of Galois theory was caused by the following question, whose answer is known as the Abel–Ruffini theorem: Why is there no formula for the roots of a fifth (or higher) degree polynomial equation in terms of the coefficients of the polynomial, using only the usual algebraic operations (addition, subtraction, multiplication, division) and application of radicals (square roots, cube roots, etc)? Galois' theory not only provides a beautiful answer to this question, but also explains in detail why it is possible to solve equations of degree four or lower in the above manner, and why their solutions take the form that they do. Further, it gives a conceptually clear, and often practical, means of telling when some particular equation of higher degree can be solved in that manner. Galois' theory also gives a clear insight into questions concerning problems in compass and straightedge construction. It gives an elegant characterization of the ratios of lengths that can be constructed with this method. Using this, it becomes relatively easy to answer such classical problems of geometry as

1. Which regular polygons are constructible polygons?
2. Why is it not possible to trisect every angle using a compass and straightedge?

History of Galois theory

Pre-history

Why is Galois theory called Galois theory? The answer is that it is named after a French mathematician Evariste Galois (1811-1832) who did some very important work in this area. He had a very dramatic and difficult life, failing to get much of his work recognised due to his great difficulty in expressing himself clearly. For example, he wasn't admitted to the leading university in Paris, the Ecole Polytechnique, and had to make do with the Ecole Normale. He also met with difficulty because of his political sympathies, he was a republican.

This led to him being expelled from the Ecole Normale when he wrote a letter to a newspaper criticising the director of the school. He joined a republican branch of the militia and was later imprisoned (twice) because of his membership. The second time whilst in prison he fell in love with the daughter of the prison physician, Stephanie-Felice du Motel and after being released died in a duel with Perscheux d'Herbinville. The reasons for the duel are not entirely clear, but it seems likely it had something to do with Stephanie. His death started republican riots and rallies which lasted for several days.

Although Galois is often credited with inventing group theory and Galois theory, it seems that an Italian mathematician Paolo Ruffini (1765-1822) may have come up with many of the ideas first. Unfortunately his ideas were not taken seriously by the rest of the mathematical community at the time. There are some links at the end of this document for anyone interested in finding out more about the history of group theory and Galois theory.

Galois' theory originated in the study of symmetric functions – the coefficients of a monic polynomial are (up to sign) the elementary symmetric polynomials in the roots. For instance, $(x - a)(x - b) = x2 - (a + b)x + ab$, where 1, $a + b$ and ab are the elementary polynomials of degree 0, 1 and 2 in two variables. This was first formalized by the 16th-century French mathematician François Viète, in Viète's formulas, for the case of positive real roots. In the opinion of the 18th-century British mathematician Charles Hutton, the expression of coefficients of a polynomial in terms of the roots (not only for positive roots) was first understood by the 17th-century French mathematician Albert Girard; Hutton writes:...[Girard was] the first person who understood the general doctrine of the formation of the coefficients of the powers from the sum of the roots and their products.

He was the first who discovered the rules for summing the powers of the roots of any equation. In this vein, the discriminant is a symmetric function in the roots that reflects properties of the roots – it is zero if and only if the polynomial has a multiple root, and for quadratic and cubic polynomials it is positive if and only if all roots are real and distinct, and negative if and only if there is a pair of distinct complex conjugate roots. See Discriminant:Nature of the roots for details. The cubic was first partly solved by the 15–16th-century Italian mathematician Scipione del Ferro, who did not however publish his results; this method, though, only solved one type of cubic equation.

This solution was then rediscovered independently in 1535 by Niccolò Fontana Tartaglia, who shared it with Gerolamo Cardano, asking him to not publish it. Cardano then extended this to numerous other cases, using similar arguments; see more details at Cardano's method. After the discovery of del Ferro's work, he felt that Tartaglia's method was no longer secret, and thus he published his solution in his 1545 Ars Magna. His student Lodovico Ferrari solved the quartic polynomial; his solution was also included in Ars Magna. In this book, however, Cardano did not provide a "general formula" for the solution of a cubic equation, as he had neither complex numbers at his disposal, nor the algebraic notation to be able to describe a general cubic equation. With the benefit of modern notation and complex numbers, the formulae in this book do work in the general case, but Cardano did not know this. It was Rafael Bombelli who managed to understand how to work with complex numbers in order to solve all forms of cubic equation.

A further step was the 1770 paper Réflexions sur la résolution algébrique des équations by the French-Italian mathematician Joseph Louis Lagrange, in his method of Lagrange resolvents, where he analyzed Cardano's and Ferrari's solution of cubics and quartics by considering them in terms of permutations of the roots, which yielded an auxiliary polynomial of lower degree, providing a unified understanding of the solutions and laying the groundwork for group theory and Galois' theory. Crucially, however, he did not consider composition of permutations.

Lagrange's method did not extend to quintic equations or higher, because the resolvent had higher degree. The quintic was almost proven to have no general solutions by radicals by Paolo Ruffini in 1799, whose key insight was to use permutation groups, not just a single permutation. His solution contained a gap, which Cauchy considered minor, though this was not patched until the work of the Norwegian mathematician Niels Henrik Abel, who published a proof in 1824, thus establishing the Abel–Ruffini theorem. While Ruffini and Abel established that the general quintic could not be solved, some particular quintics can be solved, such as (x " 1)5 = 0, and the precise criterion by which a given quintic or higher polynomial could be determined to be solvable or not was given by Évariste Galois, who showed that whether a polynomial was solvable or not was equivalent to whether or not the permutation group of its roots – in modern terms, its Galois group – had a certain structure – in modern terms, whether or not it was a solvable group. This group was always solvable for polynomials of degree four or less, but not always so for polynomials of degree five and greater, which explains why there is no general solution in higher degree.

Under the guidance of Louis Richard, one of his teachers at Louis-le-Grand, Galois's further study of algebra led him to take up the question of the solution of algebraic equations. Mathematicians for a long time had used explicit formulas, involving only rational operations and extractions of roots, for the solution of equations up to degree four, but they had been defeated by equations of degree five and higher. In 1770 Lagrange took the novel but decisive step of treating the roots of an equation as objects in their own right and studying permutations (a change in an ordered arrangement) of them. In 1799 the Italian mathematician Paolo Ruffini attempted to prove the impossibility of solving the general quintic equation by radicals. Ruffini's effort was not wholly successful, but in 1824 the Norwegian mathematician Niels Abel gave a correct proof.

Galois, stimulated by Lagrange's ideas and initially unaware of Abel's work, began searching for the necessary and sufficient conditions under which an algebraic equation of any degree can be solved by radicals. His method was to analyze the "admissible" permutations of the roots of the equation. His key discovery, brilliant and highly imaginative, was that solvability by radicals is possible if and only if the group of automorphisms (functions that take elements of a set to other elements of the set while preserving algebraic operations) is solvable, which means essentially that the group can be broken down into simple "prime-order" constituents that always have an easily understood structure. The term solvable is used because of this connection with solvability by radicals. Thus, Galois perceived that solving equations of the quintic and beyond required a wholly different kind of treatment than that required for quadratic, cubic, and quartic equations. Although Galois used the concept of group and other associated concepts, such as coset and subgroup, he did not actually define these concepts, and he did not construct a rigorous formal theory.

While still at Louis-le-Grand, Galois published one minor paper, but his life was soon overtaken by disappointment and tragedy. A memoir on the solvability of algebraic equations that he had submitted in 1829 to the French Academy of Sciences was lost by Augustin-Louis Cauchy. He failed in two attempts (1827 and

1829) to gain admission to the École Polytechnique, the leading school of French mathematics, his second attempt marred by a disastrous encounter with an oral examiner. Also in 1829 his father, after bitter clashes with conservative elements in his hometown, committed suicide. The same year, Galois enrolled as a student teacher in the less prestigious École Normale Supérieure and turned to political activism. Meanwhile he continued his research, and in the spring of 1830 he had three short articles published. At the same time, he rewrote the paper that had been lost and presented it again to the Academy—but for a second time the manuscript went astray. Jean-Baptiste-Joseph Fourier took it home but died a few weeks later, and the manuscript was never found.

The July Revolution of 1830 sent the last Bourbon monarch, Charles X, into exile. But republicans were deeply disappointed when yet another king, Louis-Philippe, ascended the throne—even though he was the "Citizen King" and wore the tricoloured flag of the French Revolution. When Galois wrote a vigorous article expressing pro-republican views, he was promptly expelled from the École Normale Supérieure. Subsequently, he was arrested twice for republican activities; he was acquitted the first time but spent six months in prison on the second charge. In 1831 he presented his memoir on the theory of equations for the third time to the Academy. This time it was returned but with a negative report. The judges, who included Siméon-Denis Poisson, did not understand what Galois had written and (incorrectly) believed that it contained a significant error. They had been quite unable to accept Galois's original ideas and revolutionary mathematical methods.

The circumstances that led to Galois's death in a duel in Paris are not altogether clear, but recent scholarship suggests that it was at his own insistence that the duel was staged and fought to look like a police ambush. In any case, anticipating his death the night before the duel, Galois hastily wrote a scientific last testament addressed to his friend Auguste Chevalier in which he summarized his work and included some new theorems and conjectures.

Galois' writings

In 1830 Galois (at the age of 18) submitted to the Paris Academy of Sciences a memoir on his theory of solvability by radicals; Galois' paper was ultimately rejected in 1831 as being too sketchy and for giving a condition in terms of the roots of the equation instead of its coefficients. Galois then died in a duel in 1832, and his paper, "Mémoire sur les conditions de résolubilité des équations par radicaux", remained unpublished until 1846 when it was published by Joseph Liouville accompanied by some of his own explanations. Prior to this publication, Liouville announced Galois' result to the Academy in a speech he gave on 4 July 1843. According to

Allan Clark, Galois's characterization "dramatically supersedes the work of Abel and Ruffini."

Aftermath

Galois' theory was notoriously difficult for his contemporaries to understand, especially to the level where they could expand on it. For example, in his 1846 commentary, Liouville completely missed the group-theoretic core of Galois' method.Joseph Alfred Serret who attended some of Liouville's talks, included Galois' theory in his 1866 (third edition) of his textbook Cours d'algèbre supérieure. Serret's pupil, Camille Jordan, had an even better understanding reflected in his 1870 book Traité des substitutions et des équations algébriques. Outside France, Galois' theory remained more obscure for a longer period. In Britain, Cayley failed to grasp its depth and popular British algebra textbooks did not even mention Galois' theory until well after the turn of the century. In Germany, Kronecker's writings focused more on Abel's result. Dedekind wrote little about Galois' theory, but lectured on it at Göttingen in 1858, showing a very good understanding.Eugen Netto's books of the 1880s, based on Jordan's Traité, made Galois theory accessible to a wider German and American audience as did Heinrich Martin Weber's 1895 algebra textbook.

Permutation group approach to Galois theory

Given a polynomial, it may be that some of the roots are connected by various algebraic equations. For example, it may be that for two of the roots, say A and B, that A2 + 5B3 = 7. The central idea of Galois' theory is to consider permutations (or rearrangements) of the roots such that any algebraic equation satisfied by the roots is still satisfied after the roots have been permuted. Originally, the theory has been developed for algebraic equations whose coefficients are rational numbers. It extends naturally to equations with coefficients in any field, but this will not be considered in the simple examples below. These permutations together form a permutation group, also called the Galois group of the polynomial, which is explicitly described in the following examples.

First example: a quadratic equation

$$x^2 - 4x + 1 = 0.$$

By using the quadratic formula, we find that the two roots are

$$\begin{aligned} A &= 2+\sqrt{3}, \\ B &= 2-\sqrt{3}. \end{aligned}$$

Examples of algebraic equations satisfied by A and B include A+ B = 4,

And AB= 1.

Obviously, in either of these equations, if we exchange A andB, we obtain another true statement. For example, the equation A + B = 4 becomes simply B + A = 4. Furthermore, it is true, but less obvious, that this holds for every possible algebraic relation between A and B such that all coefficients are rational (in any such relation, swapping A and B yields another true relation). This results from the theory of symmetric polynomials, which, in this simple case, may be replaced by formula manipulations involving binomial theorem. (One might object that A and B are related by the algebraic equation A - B - 2 $\sqrt{3}$ = 0, which does not remain true when A andB are exchanged. However, this relation is not considered here, because it has the coefficient - 2 $\sqrt{3}$ which is not rational.)

We conclude that the Galois group of the polynomial x2 " 4x + 1 consists of two permutations: the identity permutation which leaves A and B untouched, and the transposition permutation which exchanges A and B. It is a cyclic group of order two, and therefore isomorphic to Z/2Z. A similar discussion applies to any quadratic polynomial ax2 + bx + c, where a, b and c are rational numbers.

- If the polynomial has rational roots, for then the Galois group is trivial; examplex^2 - 4x + 4 = $(x-2)^2$, or x^2 - 3x +2 = (x-2) (x-1), that is, it contains only the identity permutation.
- If it has two irrational roots, for example x2 - 2, then the Galois group contains two permutations, just as in the above example.

SECOND EXAMPLE

Consider the polynomial x^4 -10 x^2 + 1,

which can also be written as $(x^2 - 5)^2 - 24$

We wish to describe the Galois group of this polynomial, again over the field of rational numbers. The polynomial has four roots:

$$\begin{aligned} A &= \sqrt{2}+\sqrt{3}, \\ B &= \sqrt{2}-\sqrt{3}, \\ C &= -\sqrt{2}+\sqrt{3}, \\ D &= -\sqrt{2}-\sqrt{3}. \end{aligned}$$

There are 24 possible ways to permute these four roots, but not all of these permutations are members of the Galois group. The members of the Galois group must preserve any algebraic equation with rational coefficients involving A, B, C and D.

Among these equations, we have:

$$\begin{aligned} AB &= -1 \\ AC &= 1 \\ A+D &= 0 \end{aligned}$$

It follows that, if ö is a permutation that belongs to the Galois group, we must have:

$$\begin{aligned} \varphi(B) &= \frac{-1}{\varphi(A)}, \\ \varphi(C) &= \frac{1}{\varphi(A)}, \\ \varphi(D) &= -\varphi(A). \end{aligned}$$

This implies that the permutation is well defined by the image of A, and that the Galois group has 4 elements, which are:

(A, B, C, D) '! (A, B, C, D)

(A, B, C, D) '! (B, A, D, C)

(A, B, C, D) '! (C, D, A, B)

(A, B, C, D) '! (D, C, B, A)

This implies that the Galois group is isomorphic to the Klein four-group.

Modern approach by field theory

In the modern approach, one starts with a field extensionL/K (read "L over K"), and examines the group of field automorphisms of L/K (these are bijectivering homomorphismsá : L → L such that á(x) = x for all x"K). See the article on Galois groups for further explanation and examples The connection between the two approaches is as follows. The coefficients of the polynomial in question should be chosen from the base field K. The top field L should be the field obtained by adjoining the roots of the polynomial in question to the base field. Any permutation of the roots which respects algebraic equations as described above gives rise to an automorphism of L/K, and vice versa. In the first example above, we were studying the extension

Q (√3)Q where Q is the field of rational numbers, and

Q (√3) is the field obtained from Q by adjoining In the second example, we were studying the extension Q(A,B,C,D)/Q. There are several advantages to the modern approach over the permutation group approach.

- It permits a far simpler statement of the fundamental theorem of Galois theory.
- The use of base fields other than Q is crucial in many areas of mathematics. For example, in algebraic number theory, one often does Galois theory using number fields, finite fields or local fields as the base field.
- It allows one to more easily study infinite extensions. Again this is important in algebraic number theory, where for example one often discusses the absolute Galois group of Q, defined to be the Galois group of K/Q where K is an algebraic closure of Q.
- It allows for consideration of inseparable extensions. This issue does not arise in the classical framework, since it was always implicitly assumed that arithmetic took place in characteristic zero, but nonzero characteristic arises frequently in number theory and in algebraic geometry.
- It removes the rather artificial reliance on chasing roots of polynomials. That is, different polynomials may yield the same extension fields, and the modern approach recognizes the connection between these polynomials.

Solvable groups and solution by radicals

The notion of a solvable group in group theory allows one to determine whether a polynomial is solvable in radicals, depending on whether its Galois group has the property of solvability. In essence, each field extension L/K corresponds to a factor group in a composition series of the Galois group. If a factor group in the composition series is cyclic of order n, and if in the corresponding field extension L/K the field K already contains a primitive nth root of unity, then it is a radical extension and the elements of L can then be expressed using the nth root of some element of K.

If all the factor groups in its composition series are cyclic, the Galois group is called solvable, and all of the elements of the corresponding field can be found by repeatedly taking roots, products, and sums of elements from the base field (usually Q). One of the great triumphs of Galois Theory was the proof that for every n> 4, there exist polynomials of degree n which are not solvable by radicals (this was proven independently, using a similar method, by Niels Henrik Abel a few years before, and is the Abel–Ruffini theorem), and a systematic way for testing whether a specific polynomial is solvable by radicals. The Abel–Ruffini theorem results from the fact that for n> 4 the symmetric groupSn contains a simple, noncyclic, normal subgroup, namely the alternating groupAn.

A non-solvable quintic example

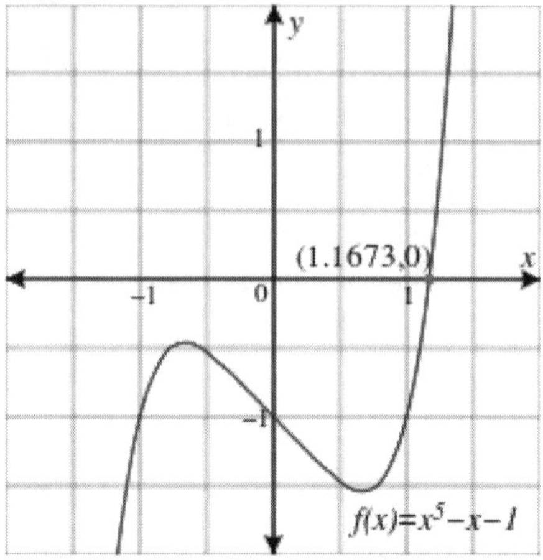

For the polynomial f(x) = x5 Š x Š 1, the lone real root x = 1.1673… is algebraic, but not expressible in terms of radicals. The other four roots are complex numbers. Van der Waerden cites the polynomial f(x) = x5 - x - 1. By the rational root theorem this has no rational zeroes. Neither does it have linear factors modulo 2 or 3. The Galois group of f(x) modulo 2 is cyclic of order 6, because f(x) modulo 2 factors into polynomials of orders 2 and 3, (x2 + x + 1)(x3 + x2 + 1). f(x) modulo 3 has no linear or quadratic factor, and hence is irreducible. Thus its modulo 3 Galois group contains an element of order 5. It is known that a Galois group modulo a prime is isomorphic to a subgroup of the Galois group over the rationals. A permutation group on 5 objects with elements of orders 6 and 5 must be the symmetric group S5, which is therefore the Galois group of f(x). This is one of the simplest examples of a non-solvable quintic polynomial. According to Serge Lang, Emil Artin found this example.

Inverse Galois problem

The inverse Galois problem is to find a field extension with a given Galois group As long as one does not also specify the ground field, the problem is not very difficult, and all finite groups do occur as Galois groups. For showing this, one may proceed as follows. Choose a field K and a finite group G. Cayley's theorem says that G is (up to isomorphism) a subgroup of the symmetric groupS on the elements of G. Choose indeterminates {xá}, one for each element á of G, and adjoin them to K to get the field F = K({xá}). Contained within F is the field L of symmetric rational functions in the {xá}. The Galois group of F/L is S, by a basic result of Emil Artin. G acts on F by restriction of action of S. If the fixed field of this action is M, then, by the fundamental theorem of Galois theory, the Galois group of F/M is G.

On the other hand, it is an open problem whether every finite group is the Galois group of a field extension of the field Q of the rational numbers. Igor Shafarevich proved that every solvable finite group is the Galois group of some extension of Q. Various people have solved the inverse Galois problem for selected non-Abelian simple groups. Existence of solutions has been shown for all but possibly one (Mathieu groupM23) of the 26 sporadic simple groups. There is even a polynomial with integral coefficients whose Galois group is the Monster group.

Inverse Galois problem

In Galois theory, the inverse Galois problem concerns whether or not every finite group appears as the Galois group of some Galois extension of the rational numbers Q. This problem, first posed in the early 19th century, is unsolved. There are some permutation groups for which generic polynomials are known, which define all algebraic extensions of Q having a particular group as Galois group. These groups include all of degree no greater than 5. There also are groups known not to have generic polynomials, such as the cyclic group of order 8. More generally, let G be a given finite group, and let K be a field. Then the question is this: is there a Galois extension field L/K such that the Galois group of the extension is isomorphic to G? One says that G is realizable over K if such a field L exists.

Partial results

There is a great deal of detailed information in particular cases. It is known that every finite group is realizable over any function field in one variable over the complex numbers C, and more generally over function fields in one variable over any algebraically closed field of characteristic zero. Igor Shafarevich showed that every finite solvable group is realizable over Q. It is also known that every sporadic group, except possibly the Mathieu group M23, is realizable over Q.

David Hilbert had shown that this question is related to a rationality question for G: If K is any extension of Q, on which G acts as an automorphism group and the invariant field KG is rational over Q, then G is realizable over Q. Here rational means that it is a purely transcendental extension of Q, generated by an algebraically independent set. This criterion can for example be used to show that all the symmetric groups are realizable. Much detailed work has been carried out on the question, which is in no sense solved in general. Some of this is based on constructing G geometrically as a Galois covering of the projective line: in algebraic terms, starting with an extension of the field Q(t) of rational functions in an indeterminate t. After that, one applies Hilbert's irreducibility theorem to specialise t, in such a way as to preserve the Galois group. All permutation groups of degree

16 or less are known to be realizable over Q ; the group PSL(2,16):2 of degree 17 may not be. All 13 non-Abelian simple groups smaller than PSL(2,25) (order 7800) are known to be realizable over Q.

A SIMPLE EXAMPLE: CYCLIC GROUPS

It is possible, using classical results, to construct explicitly a polynomial whose Galois group over Q is the cyclic group Z/nZ for any positive integer n. To do this, choose a prime p such that p a ≡1 (mod n); this is possible by Dirichlet's theorem. Let Q(ì) be the cyclotomic extension of Q generated by ì, where ì is a primitive pth root of unity; the Galois group of Q(ì)/Q is cyclic of order p - 1. Since n divides p Š 1, the Galois group has a cyclic subgroup H of order (p - 1)/n. The fundamental theorem of Galois theory implies that the corresponding fixed field, F = Q(ì)H, has Galois group Z/nZ over Q. By taking appropriate sums of conjugates of ì, following the construction of Gaussian periods, one can find an element á of F that generates F over Q, and compute its minimal polynomial. This method can be extended to cover all finite abelian groups, since every such group appears in fact as a quotient of the Galois group of some cyclotomic extension of Q. (This statement should not though be confused with the Kronecker–Weber theorem, which lies significantly deeper.)

2Worked example: the cyclic group of order three

For n = 3, we may take p = 7. Then Gal(Q(ì)/Q) is cyclic of order six. Let us take the generator ç of this group which sends ì to ì3. We are interested in the subgroup H = {1, ç3} of order two. Consider the element á = ì + ç3(ì). By construction, á is fixed by H, and only has three conjugates over Q:

$$\text{á} = \text{ç}^{0}(\text{á}) = \text{ì} + \text{ì}^{6},$$

$$\text{â} = \text{ç}^{1}(\text{á}) = \text{ì}^{3} + \text{ì}^{4},$$

$$\text{ã} = \text{ç}^{2}(\text{á}) = \text{ì}^{2} + \text{ì}^{5}.$$

Using the identity:

$$1 + \text{ì} + \text{ì}^{2} + ... + \text{ì}^{6} = 0,$$

one finds that

$$\text{á} + \text{â} + \text{ã} = \text{“}1,$$

$$\text{áâ} + \text{âã} + \text{ãá} = \text{“}2,$$

$$\text{áâã} = 1.$$

Therefore á is a root of the polynomial

$$(x \text{ “ á})(x \text{ “ â})(x \text{ “ ã}) = x^{3} + x^{2} \text{ “ } 2x \text{ “ } 1,$$

which consequently has Galois group Z/3Z over Q.

Symmetric and alternating groups

Hilbert showed that all symmetric and alternating groups are represented as Galois groups of polynomials with rational coefficients. The polynomial xn + ax + b has discriminant

We take the special case

f(x, s) = xn “ sx “ s.

Substituting a prime integer for s in f(x, s) gives a polynomial (called a specialization of f(x, s)) that by Eisenstein's criterion is irreducible. Then f(x, s) must be irreducible over Q(s). Furthermore, f(x, s) can be written and f(x, 1/2) can be factored to: whose second factor is irreducible by Eisenstein's criterion. We have now shown that the group Gal(f(x, s)/Q(s)) is doubly transitive. We can then find that this Galois group has a transposition. Use the scaling (1 Š n)x = ny to get and with we arrive at:

g(y, t) = yn “ nty + (n “ 1)t

which can be arranged to

yn “ y “ (n “ 1)(y “ 1) + (t “ 1)(“ny + n “ 1).

Gal(f(x, s)/Q(s)) is implied. Any finite doubly transitive permutation group containing a transposition is a full symmetric group. Hilbert's irreducibility theorem then implies that an infinite set of rational numbers give specializations of f(x, t) whose Galois groups are Sn over the rational field Q. In fact this set of rational numbers is dense in Q.

The discriminant of g(y, t) equals and this is not in general a perfect square. Alternating groupsSolutions for alternating groups must be handled differently for odd and even degrees. Odd DegreeLet Under this substitution the discriminant of g(y, t) equals which is a perfect square when n is odd. Even DegreeLet: Under this substitution the discriminant of g(y, t) equals: which is a perfect square when n is even. Again, Hilbert's irreducibility theorem implies the existence of infinitely many specializations whose Galois groups are alternating groups.

Rigid groups

Suppose that C1,..., Cn are conjugacy classes of a finite group G, and A be the set of n-tuples (g1,..., gn) of G such that gi is in Ci and the product g1...gn is trivial. Then A is called rigid if it is nonempty, G acts transitively on it by conjugation, and each element of A generates G. Thompson (1984) showed that if a finite group G has a rigid set then it can often be realized as a Galois group over a cyclotomic extension of the rationals. (More precisely, over the cyclotomic extension of the rationals generated by the values of the irreducible characters of G on the conjugacy classes Ci.)

This can be used to show that many finite simple groups, including the monster group, are Galois groups of extensions of the rationals. The monster group is generated by a triad of elements of orders 2, 3, and 29. All such triads are conjugate. The prototype for rigidity is the symmetric group Sn, which is generated by an n-cycle and a transposition whose product is an (n Š 1)-cycle. The construction in the preceding section used these generators to establish a polynomial's Galois group.

A construction with an elliptic modular function

Let n > 1 be any integer. A lattice Ë in the complex plane with period ratio ô has a sublattice Ë2 with period ratio nô. The latter lattice is one of a finite set of sublattices permuted by the modular group PSL(2, Z), which is based on changes of basis for Ë. Let j denote the elliptic modular function of Felix Klein. Define the polynomial ön as the product of the differences (X Š j(Ëi)) over the conjugate sublattices. As a polynomial in X, ön has coefficients that are polynomials over Q in j(ô). On the conjugate lattices, the modular group acts as PGL(2, Z/nZ). It follows that ön has Galois group isomorphic to PGL(2, Z/nZ) over Q(J(ô)).

Use of Hilbert's irreducibility theorem gives an infinite (and dense) set of rational numbers specializing ön to polynomials with Galois group PGL(2, Z/nZ) over Q. The groups PGL(2, Z/nZ) include infinitely many non-solvable groups.

Grothendieck's Galois theory

In mathematics, Grothendieck's Galois theory is an abstract approach to the Galois theory of fields, developed around 1960 to provide a way to study the fundamental group of algebraic topology in the setting of algebraic geometry. It provides, in the classical setting of field theory, an alternative perspective to that of Emil Artin based on linear algebra, which became standard from about the 1930s.

The approach of Alexander Grothendieck is concerned with the category-theoretic properties that characterise the categories of finite G-sets for a fixed profinite group G. For example, G might be the group denoted Z ^ {{\hat {\mathbb {Z}}}} {\hat {\mathbb{Z}}}, which is the inverse limit of the cyclic additive groups Z/nZ — or equivalently the completion of the infinite cyclic group Z for the topology of subgroups of finite index. A finite G-set is then a finite set X on which G acts through a quotient finite cyclic group, so that it is specified by giving some permutation of X.

In the above example, a connection with classical Galois theory can be seen by regarding Z ^ {{\hat {\mathbb {Z}}}} {\hat {\mathbb{Z}}} as the profinite Galois group Gal(F/F) of the algebraic closure F of any finite field F, over F. That is, the automorphisms of F fixing F are described by the inverse limit, as we take larger and larger finite splitting fields over F. The connection with geometry can be seen when we look at covering spaces of the unit disk in the complex plane with the origin removed: the finite covering realised by the zn map of the disk, thought of by means of a complex number variable z, corresponds to the subgroup n.Z of the fundamental group of the punctured disk.

The theory of Grothendieck, published in SGA1, shows how to reconstruct the category of G-sets from a fibre functor F, which in the geometric setting takes the fibre of a covering above a fixed base point (as a set). In fact there is an isomorphism proved of the type $G \cong \mathrm{Aut}\ (\Phi)$,

the latter being the group of automorphisms (self-natural equivalences) of F. An abstract classification of categories with a functor to the category of sets is given, by means of which one can recognise categories of G-sets for G profinite.

To see how this applies to the case of fields, one has to study the tensor product of fields. In topos theory this is a part of the study of atomic toposes.

9

IMPORTANCE OF REPRESENTATION THEORY

Representation theory emerged in the late 19th century, but many mathematicians, include prominent William Burnside, deemed it useless and redundant. In just a bit over a century following its debut, it has transformed mathematics drastically.

Representation theory is a mathematical method described precisely by what representation means: taking simpler objects to represent more complicated objects. Typically, these objects are collections of elements that are closely related to each other. If a collection A represents B, most likely any relations between elements a1 and a2 would be preserved between b1 and b2 if ai "represents" bi. The more complex objects are called groups, an integral part of abstract algebra, and the simpler ones are collections of matrices, important structures in linear algebra.

Representation theory is a branch of mathematics that studies abstract algebraic structures by representing their elements as linear transformations of vector spaces, and studies modules over these abstract algebraic structures. In essence, a representation makes an abstract algebraic object more concrete by describing its elements by matrices and their algebraic operations (for example, matrix addition, matrix multiplication). The theory of matrices and linear operators is well-understood, so representations of more abstract objects in terms of familiar linear algebra objects helps glean properties and sometimes simplify calculations on more abstract theories.

The algebraic objects amenable to such a description include groups, associative algebras and Lie algebras. The most prominent of these (and historically the first) is the representation theory of groups, in which elements of a group are represented by invertible matrices in such a way that the group operation is matrix multiplication.

Representation theory is a useful method because it reduces problems in abstract algebra to problems in linear algebra, a subject that is well understood. Furthermore, the vector space on which a group (for example) is represented can be infinite-dimensional, and by allowing it to be, for instance, a Hilbert space, methods of analysis can be applied to the theory of groups. Representation theory is also important in physics because, for example, it describes how the symmetry group of a physical system affects the solutions of equations describing that system.

Representation theory is pervasive across fields of mathematics for two reasons. First, the applications of representation theory are diverse in addition to its impact on algebra, representation theory:

illuminates and generalizes Fourier analysis via harmonic analysis,

is connected to geometry via invariant theory and the Erlangen program,

has an impact in number theory via automorphic forms and the Langlands program.

Second, there are diverse approaches to representation theory. The same objects can be studied using methods from algebraic geometry, module theory, analytic number theory, differential geometry, operator theory, algebraic combinatorics and topology.

The success of representation theory has led to numerous generalizations. One of the most general is in category theory. The algebraic objects to which representation theory applies can be viewed as particular kinds of categories, and the representations as functors from the object category to the category of vector spaces. This description points to two obvious generalizations: first, the algebraic objects can be replaced by more general categories; second, the target category of vector spaces can be replaced by other well-understood categories.

DEFINITIONS AND CONCEPTS

Let V be a vector space over a field F. For instance, suppose V is Rn or Cn, the standard n-dimensional space of column vectors over the real or complex numbers respectively. In this case, the idea of representation theory is to do abstract algebra concretely by using n × n matrices of real or complex numbers. There are three main sorts of algebraic objects for which this can be done: groups, associative algebras and Lie algebras.

- The set of all invertible n × n matrices is a group under matrix multiplication and the representation theory of groups analyzes a group by describing ("representing") its elements in terms of invertible matrices.

- Matrix addition and multiplication make the set of all n × n matrices into an associative algebra and hence there is a corresponding representation theory of associative algebras.
- If we replace matrix multiplication MN by the matrix commutator MN " NM, then the n × n matrices become instead a Lie algebra, leading to a representation theory of Lie algebras.

This generalizes to any field F and any vector space V over F, with linear maps replacing matrices and composition replacing matrix multiplication: there is a group GL(V,F) of automorphisms of V, an associative algebra EndF(V) of all endomorphisms of V, and a corresponding Lie algebra gl(V,F).

Definition

There are two ways to say what a representation is. The first uses the idea of an action, generalizing the way that matrices act on column vectors by matrix multiplication. A representation of a group G or (associative or Lie) algebra A on a vector space V is a map $\Phi : G\times V\to V$ or $\Phi : A\times V\to V$

with two properties. First, for any g in G (or a in A), the map

$$\begin{aligned} \ddot{O}(g):V &\to V \\ v &\mapsto \ddot{O}(g,v) \end{aligned}$$

is linear (over F). Second, if we introduce the notation g · v for Ö Φ (g, v), then for any g1, g2 in G and v in V:

where e is the identity element of G and g1g2 is the product in G. The requirement for associative algebras is analogous, except that associative algebras do not always have an identity element, in which case equation (1) is ignored. Equation (2) is an abstract expression of the associativity of matrix multiplication. This doesn't hold for the matrix commutator and also there is no identity element for the commutator. Hence for Lie algebras, the only requirement is that for any x1, x2 in A and v in V: where [x1, x2] is the Lie bracket, which generalizes the matrix commutator MN " NM. The second way to define a representation focuses on the map ö sending g in G to a linear map ö(g): V '! V, which satisfies

$$\ddot{o}(g_1g_2)=\ddot{o}(g_1)\circ\ddot{o}(g_2) \quad \text{for all } g_1,g_2\in G$$

and similarly in the other cases. This approach is both more concise and more abstract. From this point of view:

- A representation of a group G on a vector space V is a group homomorphism ö: G '! GL(V,F);

- A representation of an associative algebra A on a vector space V is an algebra homomorphism ö: A ’! EndF(V);
- A representation of a Lie algebra a on a vector space V is a Lie algebra homomorphism ö: a ’! gl(V,F).

Terminology

The vector space V is called the representation space of ö and its dimension (if finite) is called the dimension of the representation (sometimes degree, as in). It is also common practice to refer to V itself as the representation when the homomorphism ö is clear from the context; otherwise the notation (V,ö) can be used to denote a representation. When V is of finite dimension n, one can choose a basis for V to identify V with Fn and hence recover a matrix representation with entries in the field F. An effective or faithful representation is a representation (V,ö) for which the homomorphism ö is injective.

Equivariant maps and isomorphisms

If V and W are vector spaces over F, equipped with representations ö and ø of a group G, then an equivariant map from V to W is a linear map á: V ’! W such that

a$(g.v) = g.a(v)$

for all g in G and v in V. In terms of ö: G ’! GL(V) and ø: G ’! GL(W), this means

a o $\phi(g) = \psi(g)$ o a

for all g in G, that is, the following diagram commutes:

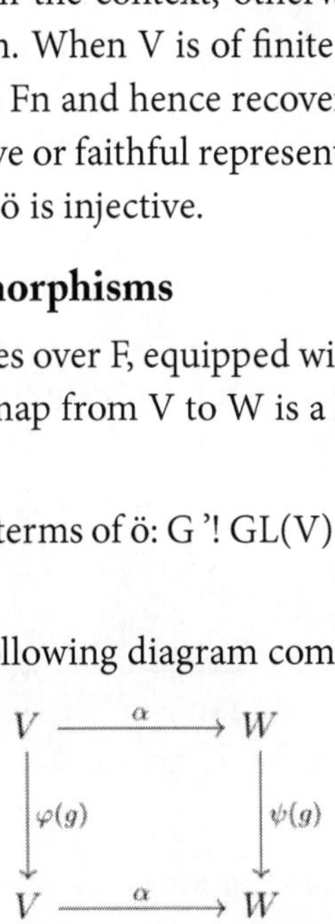

Equivariant maps for representations of an associative or Lie algebra are defined similarly. If á is invertible, then it is said to be an isomorphism, in which case V and W (or, more precisely, ö and ø) are isomorphic representations, also phrased as equivalent representations. An equivariant map is often called an intertwining map of representations. Also, in the case of a group G, it is on occasion called a G-map.

Isomorphic representations are, for practical purposes, “the same”; they provide the same information about the group or algebra being represented. Representation theory therefore seeks to classify representations up to isomorphism. Subrepresentations, quotients, and irreducible representations If (V,ø) is a representation of (say) a group G, and W is a linear subspace of V that is preserved by the action of G in the sense that g · w " W for all w " W (Serre calls these W stable under G), then W is called a subrepresentation: by defining ö(g) to be the restriction of ø(g) to W, (W, ö) is a representation of G and the inclusion of W into V is an equivariant map. The quotient space V/W can also be made into a representation of G.

If V has exactly two subrepresentations, namely the trivial subspace {0} and V itself, then the representation is said to be irreducible; if V has a proper nontrivial subrepresentation, the representation is said to be reducible. The definition of an irreducible representation implies Schur's lemma: an equivariant map á: V '! W between irreducible representations is either the zero map or an isomorphism, since its kernel and image are subrepresentations. In particular, when V = W, this shows that the equivariant endomorphisms of V form an associative division algebra over the underlying field F.

If F is algebraically closed, the only equivariant endomorphisms of an irreducible representation are the scalar multiples of the identity. Irreducible representations are the building blocks of representation theory: if a representation V is not irreducible then it is built from a subrepresentation and a quotient that are both "simpler" in some sense; for instance, if V is finite-dimensional, then both the subrepresentation and the quotient have smaller dimension.

DIRECT SUMS AND INDECOMPOSABLE REPRESENTATIONS

If (V,ö) and (W,ø) are representations of (say) a group G, then the direct sum of V and W is a representation, in a canonical way, via the equation g· (v,w) = (g·v,g·w).

The direct sum of two representations carries no more information about the group G than the two representations do individually. If a representation is the direct sum of two proper nontrivial subrepresentations, it is said to be decomposable. Otherwise, it is said to be indecomposable.

Complete reducibility

In favorable circumstances, every finite-dimensional representation is a direct sum of irreducible representations: such representations are said to be semisimple. In this case, it suffices to understand only the irreducible representations. Examples where this "complete reducibility" phenomenon occur include finite and compact groups, and semisimple Lie algebras. In cases where complete reducibility does not hold, one must understand how indecomposable representations can be built from irreducible representations as extensions of a quotient by a subrepresentation. Representation theory is notable for the number of branches it has, and the diversity of the approaches to studying representations of groups and algebras. Although, all the theories have in common the basic concepts discussed already, they differ considerably in detail. The differences are at least 3-fold:

1. Representation theory depends upon the type of algebraic object being represented. There are several different classes of groups, associative algebras and Lie algebras, and their representation theories all have an individual flavour.

2 Representation theory depends upon the nature of the vector space on which the algebraic object is represented. The most important distinction is between finite-dimensional representations and infinite-dimensional ones. In the infinite-dimensional case, additional structures are important (for example,

3. whether or not the space is a Hilbert space, Banach space, etc.). Additional algebraic structures can also be imposed in the finite-dimensional case.

4. Representation theory depends upon the type of field over which the vector space is defined. The most important cases are the field of complex numbers, the field of real numbers, finite fields, and fields of p-adic numbers. Additional difficulties arise for fields of positive characteristic and for fields that are not algebraically closed.

Finite groups

Group representations are a very important tool in the study of finite groups. They also arise in the applications of finite group theory to geometry and crystallography. Representations of finite groups exhibit many of the features of the general theory and point the way to other branches and topics in representation theory.

Over a field of characteristic zero, the representation of a finite group G has a number of convenient properties. First, the representations of G are semisimple (completely reducible). This is a consequence of Maschke's theorem, which states that any subrepresentation V of a G-representation W has a G-invariant complement. One proof is to choose any projection ð from W to V and replace it by its average ðG defined by ðG is equivariant, and its kernel is the required complement. The finite-dimensional G-representations can be understood using character theory: the character of a representation ö: G '! GL(V) is the class function ÷ö: G '! F defined by where T r {mathrm {Tr}} is the trace. An irreducible representation of G is completely determined by its character. Maschke's theorem holds more generally for fields of positive characteristic p, such as the finite fields, as long as the prime p is coprime to the order of G. When p and |G| have a common factor, there are G-representations that are not semisimple, which are studied in a subbranch called modular representation theory.

Inner product on V in the sense that for all g in G and v, w in W. Hence any G-representation is unitary. Unitary representations are automatically semisimple, since Maschke's result can be proven by taking the orthogonal complement of a subrepresentation. When studying representations of groups that are not finite, the unitary representations provide a good generalization of the real and complex representations of a finite group. Results such as Maschke's theorem and the unitary property that rely on averaging can be generalized to more general groups by replacing the average with an integral, provided that a suitable notion of integral can be defined. This can be done for compact topological groups (including compact Lie groups), using Haar measure, and the resulting theory is known as abstract harmonic analysis. Over arbitrary fields, another class of finite groups that have a good representation theory are the finite groups of Lie type. Important examples are linear algebraic groups over finite fields.

The representation theory of linear algebraic groups and Lie groups extends these examples to infinite-dimensional groups, the latter being intimately related to Lie algebra representations. The importance of character theory for finite groups has an analogue in the theory of weights for representations of Lie groups and Lie algebras. Representations of a finite group G are also linked directly to algebra representations via the group algebra F[G], which is a vector space over F with the elements of G as a basis, equipped with the multiplication operation defined by the group operation, linearity, and the requirement that the group operation and scalar multiplication commute.

Modular representations

Modular representations of a finite group G are representations over a field whose characteristic is not coprime to |G|, so that Maschke's theorem no longer holds (because |G| is not invertible in F and so one cannot divide by it). Nevertheless, Richard Brauer extended much of character theory to modular representations, and this theory played an important role in early progress towards the classification of finite simple groups, especially for simple groups whose characterization was not amenable to purely group-theoretic methods because their Sylow 2-subgroups were "too small". As well as having applications to group theory, modular representations arise naturally in other branches of mathematics, such as algebraic geometry, coding theory, combinatorics and number theory.

Modular representation theory

Modular representation theory is a branch of mathematics, and that part of representation theory that studies linear representations of finite groups over a field K of positive characteristic p, necessarily a prime number. As well as having applications to group theory, modular representations arise naturally in other branches of mathematics, such as algebraic geometry, coding theory, combinatorics and number theory.

Within finite group theory, character-theoretic results proved by Richard Brauer using modular representation theory played an important role in early progress towards the classification of finite simple groups, especially for simple groups whose characterization was not amenable to purely group-theoretic methods because their Sylow 2 subgroups were too small in an appropriate sense.

Also, a general result on embedding of elements of order 2 in finite groups called the Z* theorem, proved by George Glauberman using the theory developed by Brauer, was particularly useful in the classification program. If the characteristic p of K does not divide the order |G|, then modular representations are completely reducible, as with ordinary (characteristic 0) representations, by virtue of Maschke's theorem. In the other case, when |G| a" 0 mod p, the process of averaging over the group needed to prove Maschke's theorem breaks down, and representations need not be completely reducible. Much of the discussion below implicitly assumes that the field K is sufficiently large (for example, K algebraically closed suffices), otherwise some statements need refinement.

HISTORY

The earliest work on representation theory over finite fields is by Dickson (1902) who showed that when p does not divide the order of the group, the representation theory is similar to that in characteristic 0. He also investigated modular invariants of some finite groups. The systematic study of modular representations, when the characteristic p divides the order of the group, was started by Brauer (1935) and was continued by him for the next few decades.

Example

Finding a representation of the cyclic group of two elements over F2 is equivalent to the problem of finding matrices whose square is the identity matrix. Over every field of characteristic other than 2, there is always a basis such that the matrix can be written as a diagonal matrix with only 1 or "1 occurring on the diagonal, such as Over F2, there are many other possible matrices, such as Over an algebraically closed field of positive characteristic, the representation theory of a finite cyclic group is fully explained by the theory of the Jordan normal form. Non-diagonal Jordan forms occur when the characteristic divides the order of the group.

Ring theory interpretation

Given a field K and a finite group G, the group algebra K[G] (which is the K-vector space with K-basis consisting of the elements of G, endowed with algebra multiplication by extending the multiplication of G by linearity) is an Artinian ring. When the order of G is divisible by the characteristic of K, the group algebra is not semisimple, hence has non-zero Jacobson radical. In that case, there are finite-dimensional modules for the group algebra that are not projective modules. By contrast, in the characteristic 0 case every irreducible representation is a direct summand of the regular representation, hence is projective.

Brauer characters

Modular representation theory was developed by Richard Brauer from about 1940 onwards to study in greater depth the relationships between the characteristic p representation theory, ordinary character theory and structure of G, especially as the latter relates to the embedding of, and relationships between, its p-subgroups. Such results can be applied in group theory to problems not directly phrased in terms of representations.

Brauer introduced the notion now known as the Brauer character. When K is algebraically closed of positive characteristic p, there is a bijection between roots of unity in K and complex roots of unity of order prime to p. Once a choice of such a bijection is fixed, the Brauer character of a representation assigns to each group element of order coprime to p the sum of complex roots of unity corresponding to the eigenvalues (including multiplicities) of that element in the given representation.

The Brauer character of a representation determines its composition factors but not, in general, its equivalence type. The irreducible Brauer characters are those afforded by the simple modules. These are integral (though not necessarily non-negative) combinations of the restrictions to elements of order coprime to p of the ordinary irreducible characters. Conversely, the restriction to the elements of order prime to p of each ordinary irreducible character is uniquely expressible as a non-negative integer combination of irreducible Brauer characters.

Reduction (mod p)

In the theory initially developed by Brauer, the link between ordinary representation theory and modular representation theory is best exemplified by considering the group algebra of the group G over a complete discrete valuation ring R with residue field K of positive characteristic p and field of fractions F of characteristic 0, such as the p-adic integers. The structure of R[G] is closely related both to the structure of the group algebra K[G] and to the structure of the semisimple group algebra F[G], and there is much interplay between the module theory of the three algebras. Each R[G]-module naturally gives rise to an F[G]-module, and, by a process often known informally as reduction (mod p), to a K[G]-module. On the other hand, since R is a principal ideal domain, each finite-dimensional F[G]-module arises by extension of scalars from an R[G]-module. In general, however, not all K[G]-modules arise as reductions (mod p) of R[G]-modules. Those that do are liftable.

Number of simple modules

In ordinary representation theory, the number of simple modules k(G) is equal to the number of conjugacy classes of G. In the modular case, the number l(G) of simple modules is equal to the number of conjugacy classes whose elements have order coprime to the relevant prime p, the so-called p-regular classes.

Blocks and the structure of the group algebra

In modular representation theory, while Maschke's theorem does not hold when the characteristic divides the group order, the group algebra may be decomposed as the direct sum of a maximal collection of two-sided ideals known as blocks. When the field F has characteristic 0, or characteristic coprime to the group order, there is still such a decomposition of the group algebra F[G] as a sum of blocks (one for each isomorphism type of simple module), but the situation is relatively transparent when F is sufficiently large: each block is a full matrix algebra over F, the endomorphism ring of the vector space underlying the associated simple module.

To obtain the blocks, the identity element of the group G is decomposed as a sum of primitive idempotents in Z(R[G]), the center of the group algebra over the maximal order R of F. The block corresponding to the primitive idempotent e is the two-sided ideal e R[G]. For each indecomposable R[G]-module, there is only one such primitive idempotent that does not annihilate it, and the module is said to belong to (or to be in) the corresponding block (in which case, all its composition factors also belong to that block). In particular, each simple module belongs to a unique block. Each ordinary irreducible character may also be assigned to a unique block according to its decomposition as a sum of irreducible Brauer characters. The block containing the trivial module is known as the principal block.

Projective modules

In ordinary representation theory, every indecomposable module is irreducible, and so every module is projective. However, the simple modules with characteristic dividing the group order are rarely projective. Indeed, if a simple module is projective, then it is the only simple module in its block, which is then isomorphic to the endomorphism algebra of the underlying vector space, a full matrix algebra. In that case, the block is said to have 'defect 0'. Generally, the structure of projective modules is difficult to determine.

For the group algebra of a finite group, the (isomorphism types of) projective indecomposable modules are in a one-to-one correspondence with the (isomorphism types of) simple modules: the socle of each projective indecomposable is simple (and isomorphic to the top), and this affords the bijection, as non-isomorphic projective indecomposables have non-isomorphic socles. The multiplicity of a projective indecomposable module as a summand of the group algebra (viewed as the regular module) is the dimension of its socle (for large enough fields of characteristic zero, this recovers the fact that each simple module occurs with multiplicity equal to its dimension as a direct summand of the regular module).

Each projective indecomposable module (and hence each projective module) in positive characteristic p may be lifted to a module in characteristic 0. Using the ring R as above, with residue field K, the identity element of G may be decomposed as a sum of mutually orthogonal primitive idempotents (not necessarily central) of K[G]. Each projective indecomposable K[G]-module is isomorphic to e.K[G] for a primitive idempotent e that occurs in this decomposition. The idempotent e lifts to a primitive idempotent, say E, of R[G], and the left module E.R[G] has reduction (mod p) isomorphic to e.K[G].

Some orthogonality relations for Brauer characters

When a projective module is lifted, the associated character vanishes on all elements of order divisible by p, and (with consistent choice of roots of unity), agrees with the Brauer character of the original characteristic p module on p-regular elements. The (usual character-ring) inner product of the Brauer character of a projective indecomposable with any other Brauer character can thus be defined: this is 0 if the second Brauer character is that of the socle of a non-isomorphic projective indecomposable, and 1 if the second Brauer character is that of its own socle. The multiplicity of an ordinary irreducible character in the character of the lift of a projective indecomposable is equal to the number of occurrences of the Brauer character of the socle of the projective indecomposable when the restriction of the ordinary character to p-regular elements is expressed as a sum of irreducible Brauer characters.

Decomposition matrix and Cartan matrix

The composition factors of the projective indecomposable modules may be calculated as follows: Given the ordinary irreducible and irreducible Brauer characters of a particular finite group, the irreducible ordinary characters may be decomposed as non-negative integer combinations of the irreducible Brauer characters. The integers involved can be placed in a matrix, with the ordinary irreducible characters assigned rows and the irreducible Brauer characters assigned columns. This is referred to as the decomposition matrix, and is frequently labelled D. It is customary to place the trivial ordinary and Brauer characters in the first row and column respectively. The product of the transpose of D with D itself results in the Cartan matrix, usually denoted C; this is a symmetric matrix such that the entries in its j-th row are the multiplicities of the respective simple modules as composition factors of the j-th projective indecomposable module. The Cartan matrix is non-singular; in fact, its determinant is a power of the characteristic of K. Since a projective indecomposable module in a given block has all its composition factors in that same block, each block has its own Cartan matrix.

Defect groups

To each block B of the group algebra K[G], Brauer associated a certain p-subgroup, known as its defect group (where p is the characteristic of K). Formally, it is the largest p-subgroup D of G for which there is a Brauer correspondent of B for the subgroup D C G (D) {DC_{G}(D)}, where C G (D) {C_{G}(D)} is the centralizer of D in G. The defect group of a block is unique up to conjugacy and has a strong influence on the structure of the block. For example, if the defect group is trivial, then the block contains just one simple module, just one ordinary character, the ordinary and Brauer irreducible characters agree on elements of order prime to the relevant characteristic p, and the simple module is projective. At the other extreme, when K has characteristic p, the Sylow p-subgroup of the finite group G is a defect group for the principal block of K[G].

The order of the defect group of a block has many arithmetical characterizations related to representation theory. It is the largest invariant factor of the Cartan matrix of the block, and occurs with multiplicity one. Also, the power of p dividing the index of the defect group of a block is the greatest common divisor of the powers of p dividing the dimensions of the simple modules in that block, and this coincides with the greatest common divisor of the powers of p dividing the degrees of the ordinary irreducible characters in that block.

Other relationships between the defect group of a block and character theory include Brauer's result that if no conjugate of the p-part of a group element g is in the defect group of a given block, then each irreducible character in that block vanishes at g. This is one of many consequences of Brauer's second main theorem. The defect group of a block also has several characterizations in the more module-theoretic approach to block theory, building on the work of J. A. Green, which associates a p-subgroup known as the vertex to an indecomposable module, defined in terms of relative projectivity of the module. For example, the vertex of each indecomposable module in a block is contained (up to conjugacy) in the defect group of the block, and no proper subgroup of the defect group has that property. Brauer's first main theorem states that the number of blocks of a finite group that have a given p-subgroup as defect group is the same as the corresponding number for the normalizer in the group of that p-subgroup.

The easiest block structure to analyse with non-trivial defect group is when the latter is cyclic. Then there are only finitely many isomorphism types of indecomposable modules in the block, and the structure of the block is by now well understood, by virtue of work of Brauer, E.C. Dade, J.A.Green and J.G.Thompson, among others. In all other cases, there are infinitely many isomorphism types of indecomposable modules in the block. Blocks whose defect groups are not cyclic can be divided into two types: tame and wild. The tame blocks (which only occur for the prime 2) have as a defect group a dihedral group, semidihedral group or (generalized) quaternion group, and their structure has been broadly determined in a series of papers by Karin Erdmann. The indecomposable modules in wild blocks are extremely difficult to classify, even in principle.

Unitary representations

A unitary representation of a group G is a linear representation ö of G on a real or (usually) complex Hilbert space V such that ö(g) is a unitary operator for every g " G. Such representations have been widely applied in quantum mechanics since the 1920s, thanks in particular to the influence of Hermann Weyl, and this has inspired the development of the theory, most notably through the analysis of representations of the Poincaré group by Eugene Wigner. One of the pioneers in constructing a general theory of unitary representations (for any group G rather than just for particular groups useful in applications) was George Mackey, and an extensive theory was developed by Harish-Chandra and others in the 1950s and 1960s.

A major goal is to describe the "unitary dual", the space of irreducible unitary representations of G. The theory is most well-developed in the case that G is a locally compact (Hausdorff) topological group and the representations are strongly continuous. For G abelian, the unitary dual is just the space of characters, while for G compact, the Peter–Weyl theorem shows that the irreducible unitary representations are finite-dimensional and the unitary dual is discrete. For example, if G is the circle group S1, then the characters are given by integers, and the unitary dual is Z. For non-compact G, the question of which representations are unitary is a subtle one. Although irreducible unitary representations must be "admissible" (as Harish-Chandra modules) and it is easy to detect which admissible representations have a nondegenerate invariant sesquilinear form, it is hard to determine when this form is positive definite. An effective description of the unitary dual, even for relatively well-behaved groups such as real reductive Lie groups (discussed below), remains an important open problem in representation theory. It has been solved for many particular groups, such as SL(2,R) and the Lorentz group.

HARMONIC ANALYSIS

The duality between the circle group S1 and the integers Z, or more generally, between a torus Tn and Zn is well known in analysis as the theory of Fourier series, and the Fourier transform similarly expresses the fact that the space of characters on a real vector space is the dual vector space. Thus unitary representation theory and harmonic analysis are intimately related, and abstract harmonic analysis exploits this relationship, by developing the analysis of functions on locally compact topological groups and related spaces. A major goal is to provide a general form of the Fourier transform and the Plancherel theorem. This is done by constructing a measure on the unitary dual and an isomorphism between the regular representation of G on the space L2(G) of square integrable functions on G and its representation on the space of L2 functions on the unitary dual. Pontrjagin duality and the Peter–Weyl theorem achieve this for abelian and compact G respectively.

Another approach involves considering all unitary representations, not just the irreducible ones. These form a category, and Tannaka–Krein duality provides a way to recover a compact group from its category of unitary representations. If the group is neither abelian nor compact, no general theory is known with an analogue of the Plancherel theorem or Fourier inversion, although Alexander Grothendieck extended Tannaka–Krein duality to a relationship between linear algebraic groups and tannakian categories. Harmonic analysis has also been extended from the analysis of functions on a group G to functions on homogeneous spaces for G. The theory is particularly well developed for symmetric spaces and provides a theory of automorphic forms (discussed below).

Harmonic analysis

The harmonics of color. The harmonic-analysis chart shows how the different wavelengths interact with red light. At a difference of ë/2 (half wavelength), red is perfectly in sync with its second harmonic in the ultraviolet. All other wavelengths in the visual spectrum have less than a ë/2 difference between them, forming harmonic oscillations in the combined waves. At ë/14, the oscillations cycle every 14th wave, while at ë/8 they will cycle every 8th. The oscillations are most rapid at ë/4, cycling every 4th wave, while at ë/3 they cycle every 7th wave, and at ë/2.5 they cycle every 13th. The lower section shows how the ë/4 harmonic interacts in visible light (green and red), as photographed in an optical flat. Harmonic analysis is a branch of mathematics concerned with the representation of functions or signals as the superposition of basic waves, and the study of and generalization of the notions of Fourier series and Fourier transforms (i.e. an extended form of Fourier analysis).

In the past two centuries, it has become a vast subject with applications in areas as diverse as number theory, representation theory, signal processing, quantum mechanics, tidal analysis and neuroscience. The term "harmonics" originated as the Ancient Greek word harmonikos, meaning "skilled in music". In physical eigenvalue problems, it began to mean waves whose frequencies are integer multiples of one another, as are the frequencies of the harmonics of music notes, but the term has been generalized beyond its original meaning. The classical Fourier transform on Rn is still an area of ongoing research, particularly concerning Fourier transformation on more general objects such as tempered distributions. For instance, if we impose some requirements on a distribution f, we can attempt to translate these requirements in terms of the Fourier transform of f. The Paley–Wiener theorem is an example of this. The Paley–Wiener theorem immediately implies that if f is a nonzero distribution of compact support (these include functions of compact support), then its Fourier transform is never compactly supported. This is a very elementary form of an uncertainty principle in a harmonic-analysis setting.

Applied harmonic analysis

Bass-guitar time signal of open-string A note (55 Hz) Fourier transform of bass-guitar time signal of open-string A note (55 Hz) Many applications of harmonic analysis in science and engineering begin with the idea or hypothesis that a phenomenon or signal is composed of a sum of individual oscillatory components. Ocean tides and vibrating strings are common and simple examples. The theoretical approach is often to try to describe the system by a differential equation or system of equations to predict the essential features, including the amplitude, frequency, and phases of the oscillatory components. The specific equations depend on the field, but theories generally try to select equations that represent major principles that are applicable. The experimental approach is usually to acquire data that accurately quantifies the phenomenon. For example, in a study of tides, the experimentalist would acquire samples of water depth as a function of time at closely enough spaced intervals to see each oscillation and over a long enough duration that multiple oscillatory periods are likely included. In a study on vibrating strings, it is common for the experimentalist to acquire a sound waveform sampled at a rate at least twice that of the highest frequency expected and for a duration many times the period of the lowest frequency expected.

For example, the top signal at the right is a sound waveform of a bass guitar playing an open string corresponding to an A note with a fundamental frequency of 55 Hz. The waveform appears oscillatory, but it is more complex than a simple sine wave, indicating the presence of additional waves. The different wave components contributing to the sound can be revealed by applying a mathematical analysis technique known as the Fourier transform, the result of which is shown in the lower figure. Note that there is a prominent peak at 55 Hz, but that there are other peaks at 110 Hz, 165 Hz, and at other frequencies corresponding to integer multiples of 55 Hz. In this case, 55 Hz is identified as the fundamental frequency of the string vibration, and the integer multiples are known as harmonics.

ABSTRACT HARMONIC ANALYSIS

Lord Kelvin's harmonic analyser 1876, Hunterian Museum, Glasgow One of the most modern branches of harmonic analysis, having its roots in the mid-20th century, is analysis on topological groups. The core motivating ideas are the various Fourier transforms, which can be generalized to a transform of functions defined on Hausdorff locally compact topological groups. The theory for abelian locally compact groups is called Pontryagin duality. Harmonic analysis studies the properties of that duality and Fourier transform and attempts to extend those features to different settings, for instance, to the case of non-abelian Lie groups. For general non-abelian locally compact groups, harmonic analysis is closely related to the theory of unitary group representations. For compact groups, the Peter–Weyl theorem explains how one may get harmonics by choosing one irreducible representation out of each equivalence class of representations. This choice of harmonics enjoys some of the useful properties of the classical Fourier transform in terms of carrying convolutions to pointwise products, or otherwise showing a certain understanding of the underlying group structure.

case, representations in infinite dimensions play a crucial role.

Other branches

- Study of the eigenvalues and eigenvectors of the Laplacian on domains, manifolds, and (to a lesser extent) graphs is also considered a branch of harmonic analysis. See e.g., hearing the shape of a drum.
- Harmonic analysis on Euclidean spaces deals with properties of the Fourier transform on Rn that have no analog on general groups. For example, the fact that the Fourier transform is rotation-invariant. Decomposing the Fourier transform into its radial and spherical components leads to topics such as Bessel functions and spherical harmonics.
- Harmonic analysis on tube domains is concerned with generalizing properties of Hardy spaces to higher dimensions.

Lie groups

A Lie group is a group that is also a smooth manifold. Many classical groups of matrices over the real or complex numbers are Lie groups. Many of the groups important in physics and chemistry are Lie groups, and their representation theory is crucial to the application of group theory in those fields. The representation theory of Lie groups can be developed first by considering the compact groups, to which results of compact representation theory apply. This theory can be extended to finite-dimensional representations of semisimple Lie groups using Weyl's unitary trick: each semisimple real Lie group G has a complexification, which is a complex Lie group Gc, and this complex Lie group has a maximal compact subgroup K. The finite-dimensional representations of G closely correspond to those of K. A general Lie group is a semidirect product of a solvable Lie group and a semisimple Lie group (the Levi decomposition). The classification of representations of solvable Lie groups is intractable in general, but often easy in practical cases. Representations of semidirect products can then be analysed by means of general results called Mackey theory, which is a generalization of the methods used in Wigner's classification of representations of the Poincaré group.

Lie algebras

A Lie algebra over a field F is a vector space over F equipped with a skew-symmetric bilinear operation called the Lie bracket, which satisfies the Jacobi identity. Lie algebras arise in particular as tangent spaces to Lie groups at the identity element, leading to their interpretation as "infinitesimal symmetries". An important approach to the representation theory of Lie groups is to study the corresponding representation theory of Lie algebras, but representations of Lie algebras also have an intrinsic interest. Lie algebras, like Lie groups, have a

Levi decomposition into semisimple and solvable parts, with the representation theory of solvable Lie algebras being intractable in general. In contrast, the finite-dimensional representations of semisimple Lie algebras are completely understood, after work of Élie Cartan. A representation of a semisimple Lie algebra g is analysed by choosing a Cartan subalgebra, which is essentially a generic maximal subalgebra h of g on which the Lie bracket is zero ("abelian"). The representation of g can be decomposed into weight spaces that are eigenspaces for the action of h and the infinitesimal analogue of characters. The structure of semisimple Lie algebras then reduces the analysis of representations to easily understood combinatorics of the possible weights that can occur.

INFINITE-DIMENSIONAL LIE ALGEBRAS

There are many classes of infinite-dimensional Lie algebras whose representations have been studied. Among these, an important class are the Kac–Moody algebras. They are named after Victor Kac and Robert Moody, who independently discovered them. These algebras form a generalization of finite-dimensional semisimple Lie algebras, and share many of their combinatorial properties. This means that they have a class of representations that can be understood in the same way as representations of semisimple Lie algebras. Affine Lie algebras are a special case of Kac–Moody algebras, which have particular importance in mathematics and theoretical physics, especially conformal field theory and the theory of exactly solvable models. Kac discovered an elegant proof of certain combinatorial identities, Macdonald identities, which is based on the representation theory of affine Kac–Moody algebras.

Lie superalgebras

Lie superalgebras are generalizations of Lie algebras in which the underlying vector space has a Z2-grading, and skew-symmetry and Jacobi identity properties of the Lie bracket are modified by signs. Their representation theory is similar to the representation theory of Lie algebras.

Linear algebraic groups

Linear algebraic groups (or more generally, affine group schemes) are analogues in algebraic geometry of Lie groups, but over more general fields than just R or C. In particular, over finite fields, they give rise to finite groups of Lie type. Although linear algebraic groups have a classification that is very similar to that of Lie groups, their representation theory is rather different (and much less well understood) and requires different techniques, since the Zariski topology is relatively weak, and techniques from analysis are no longer available.

Invariant theory

Invariant theory studies actions on algebraic varieties from the point of view of their effect on functions, which form representations of the group. Classically, the theory dealt with the question of explicit description of polynomial functions that do not change, or are invariant, under the transformations from a given linear group. The modern approach analyses the decomposition of these representations into irreducibles. Invariant theory of infinite groups is inextricably linked with the development of linear algebra, especially, the theories of quadratic forms and determinants. Another subject with strong mutual influence is projective geometry, where invariant theory can be used to organize the subject, and during the 1960s, new life was breathed into the subject by David Mumford in the form of his geometric invariant theory. The representation theory of semisimple Lie groups has its roots in invariant theory and the strong links between representation theory and algebraic geometry have many parallels in differential geometry, beginning with Felix Klein's Erlangen program and Élie Cartan's connections, which place groups and symmetry at the heart of geometry. Modern developments link representation theory and invariant theory to areas as diverse as holonomy, differential operators and the theory of several complex variables.

Automorphic forms and number theory

Automorphic forms are a generalization of modular forms to more general analytic functions, perhaps of several complex variables, with similar transformation properties. The generalization involves replacing the modular group PSL2 (R) and a chosen congruence subgroup by a semisimple Lie group G and a discrete subgroup Ã. Just as modular forms can be viewed as differential forms on a quotient of the upper half space H = PSL2 (R)/SO(2), automorphic forms can be viewed as differential forms (or similar objects) on Ã\G/K, where K is (typically) a maximal compact subgroup of G. Some care is required, however, as the quotient typically has singularities. The quotient of a semisimple Lie group by a compact subgroup is a symmetric space and so the theory of automorphic forms is intimately related to harmonic analysis on symmetric spaces. Before the development of the general theory, many important special cases were worked out in detail, including the Hilbert modular forms and Siegel modular forms. Important results in the theory include the Selberg trace formula and the realization by Robert Langlands that the Riemann-Roch theorem could be applied to calculate the dimension of the space of automorphic forms. The subsequent notion of "automorphic representation" has proved of great technical value for dealing with the case that G is an algebraic group, treated as an adelic algebraic group. As a result, an entire philosophy, the

Langlands program has developed around the relation between representation and number theoretic properties of automorphic forms.

Associative algebras

In one sense, associative algebra representations generalize both representations of groups and Lie algebras. A representation of a group induces a representation of a corresponding group ring or group algebra, while representations of a Lie algebra correspond bijectively to representations of its universal enveloping algebra. However, the representation theory of general associative algebras does not have all of the nice properties of the representation theory of groups and Lie algebras.

Module theory

When considering representations of an associative algebra, one can forget the underlying field, and simply regard the associative algebra as a ring, and its representations as modules. This approach is surprisingly fruitful: many results in representation theory can be interpreted as special cases of results about modules over a ring.

Hopf algebras and quantum groups

Hopf algebras provide a way to improve the representation theory of associative algebras, while retaining the representation theory of groups and Lie algebras as special cases. In particular, the tensor product of two representations is a representation, as is the dual vector space. The Hopf algebras associated to groups have a commutative algebra structure, and so general Hopf algebras are known as quantum groups, although this term is often restricted to certain Hopf algebras arising as deformations of groups or their universal enveloping algebras. The representation theory of quantum groups has added surprising insights to the representation theory of Lie groups and Lie algebras, for instance through the crystal basis of Kashiwara.

Representation theory of Hopf algebras

In abstract algebra, a representation of a Hopf algebra is a representation of its underlying associative algebra. That is, a representation of a Hopf algebra H over a field K is a K-vector space V with an action H × V '! V usually denoted by juxtaposition (that is, the image of (h,v) is written hv). The vector space V is called an H-module.

Properties

The module structure of a representation of a Hopf algebra H is simply its structure as a module for the underlying associative algebra. The main use of considering the additional structure of a Hopf algebra is when considering all H-modules as a category. The additional structure is also used to define invariant elements of an H-module V. An element v in V is invariant under H if for all h in H, hv = å(h)v, where å is the counit of H. The subset of all invariant elements of V forms a submodule of V. Categories of representations as a motivation for Hopf algebras For an associative algebra H, the tensor product V1 —" V2 of two H-modules V1 and V2 is a vector space, but not necessarily an H-module. For the tensor product to be a functorial product operation on H-modules, there must be a linear binary operation Δ: H → H ⊗ H H such that for any v in V1 —" V2 and any h in H,

and for any v in V1 —" V2 and a and b in H,

$$Ä(ab)(v_{(1)} \otimes v_{(2)}) = (ab)v = a[b[v]] = Äa[Äb(v_{(1)} \otimes v_{(2)})] = (Äa)(Äb)(v_{(1)} \otimes v_{(2)}).$$

using sumless Sweedler's notation, which is somewhat like an index free form of Einstein's summation convention. This is satisfied if there is a Ä such that Ä(ab) = Ä(a)Ä(b) for all a, b in H. For the category of H-modules to be a strict monoidal category with respect to$V_1 \otimes (V_2 \otimes V_3)$ ", and $(V_1 \otimes V_2) \otimes V_3$ must be equivalent and there must be unit object $å_H$, called the trivial module, such that $å_H$ —" V, V and V —" $å_H$ are equivalent.

This means that for any v in $V_1 \otimes (V_2 \otimes V_3) = (V_1 \otimes V_2) \otimes V_3$ $V_1 \otimes (V_2 \otimes V_3) = (V_1 \otimes V_2) \otimes V_3$

and for h in H,

$$((\mathrm{id} \otimes \Delta)\Delta h)(v_{(1)} \otimes v_{(2)} \otimes v_{(3)}) = h_{(1)}v_{(1)} \otimes h_{(2)(1)}v_{(2)} \otimes h_{(2)(2)}v_{(3)} = hv = ((\Delta \otimes \mathrm{id})\Delta h)(v_{(1)} \otimes v_{(2)} \otimes v_{(3)}).$$

This will hold for any three H-modules if Ä satisfies

The trivial module must be one-dimensional, and so an algebra homomorphism å : H '! F may be defined such that hv = å(h)v for all v in åH. The trivial module may be identified with F, with 1 being the element such that 1 —" v = v = v —" 1 for all v. It follows that for any v in any H-module V, any c in åH and any h in H,

$$(å(h_{(1)})h_{(2)})cv = h_{(1)}c \otimes h_{(2)}v = h(c \otimes v) = h(cv) = (h_{(1)}å(h_{(2)}))cv.$$

The existence of an algebra homomorphism å satisfying

$$å(h_{(1)})h_{(2)} = h = h_{(1)}å(h_{(2)})$$

is a sufficient condition for the existence of the trivial module. It follows that in order for the category of H-modules to be a monoidal category with respect to the tensor product, it is sufficient for H to have maps Ä and å satisfying these conditions. This is the motivation for the definition of a bialgebra, where Ä is called the comultiplication and å is called the counit. In order for each H-module V to have a dual representation V such that the underlying vector spaces are dual and the operation * is functorial over the monoidal category of H- modules, there must be a linear map S : H '! H such that for any h in H, x in V and y in V*, where è' Å", Å" é' {langle \cdot,\cdot \rangle} is the usual pairing of dual vector spaces. If the map ö : V —" V " '! å H {varphi :V\otimes V^{*}\rightarrow \varepsilon _{H}} induced by the pairing is to be an H-homomorphism, then for any h in H, x in V and y in V*,

$$\varphi(h(x\otimes y))=\varphi(x\otimes S(h_{(1)})h_{(2)}y)=\varphi(S(h_{(2)})h_{(1)}x\otimes y)=h\varphi(x\otimes y)=\varepsilon(h)\varphi(x\otimes y),$$

which is satisfied if $S(h_{(1)})h_2 = \varepsilon(h)=h_{(1)} S(h_{(2)})$

for all h in H.

If there is such a map S, then it is called an antipode, and H is a Hopf algebra. The desire for a monoidal category of modules with functorial tensor products and dual representations is therefore one motivation for the concept of a Hopf algebra.

Representations on an algebra

A Hopf algebra also has representations which carry additional structure, namely they are algebras. Let H be a Hopf algebra. If A is an algebra with the product operation ì : A —" A '! A, and ñ : H —" A '! A is a representation of H on A, then ñ is said to be a representation of H on an algebra if ì is H-equivariant. As special cases, Lie algebras, Lie superalgebras and groups can also have representations on an algebra.

Representations of categories

Since groups are categories, one can also consider representation of other categories. The simplest generalization is to monoids, which are categories with one object. Groups are monoids for which every morphism is invertible. General monoids have representations in any category. In the category of sets, these are monoid actions, but monoid representations on vector spaces and other objects can be studied.

More generally, one can relax the assumption that the category being represented has only one object. In full generality, this is simply the theory of functors between categories, and little can be said.

One special case has had a significant impact on representation theory, namely the representation theory of quivers. A quiver is simply a directed graph (with loops and multiple arrows allowed), but it can be made into a category (and also an algebra) by considering paths in the graph. Representations of such categories/ algebras have illuminated several aspects of representation theory, for instance by allowing non-semisimple representation theory questions about a group to be reduced in some cases to semisimple representation theory questions about a quiver.

Group representation

In the mathematical field of representation theory, group representations describe abstract groups in terms of bijective linear transformations of a vector space to itself (i.e. vector space automorphisms); in particular, they can be used to represent group elements as invertible matrices so that the group operation can be represented by matrix multiplication. Representations of groups are important because they allow many group-theoretic problems to be reduced to problems in linear algebra, which is well understood. They are also important in physics because, for example, they describe how the symmetry group of a physical system affects the solutions of equations describing that system.

The term representation of a group is also used in a more general sense to mean any "description" of a group as a group of transformations of some mathematical object. More formally, a "representation" means a homomorphism from the group to the automorphism group of an object. If the object is a vector space we have a linear representation. Some people use realization for the general notion and reserve the term representation for the special case of linear representations. The bulk of this article describes linear representation theory; see the last section for generalizations.

Branches of group representation theory

The representation theory of groups divides into subtheories depending on the kind of group being represented. The various theories are quite different in detail, though some basic definitions and concepts are similar. The most important divisions are:

Finite groups — Group representations are a very important tool in the study of finite groups. They also arise in the applications of finite group theory to crystallography and to geometry. If the field of scalars of the vector space has characteristic p, and if p divides the order of the group, then this is called modular representation theory; this special case has very different properties. See Representation theory of finite groups.

Compact groups or locally compact groups — Many of the results of finite group representation theory are proved by averaging over the group. These proofs can be carried over to infinite groups by replacement of the average with an integral, provided that an acceptable notion of integral can be defined. This can be done for locally compact groups, using Haar measure. The resulting theory is a central part of harmonic analysis. The Pontryagin duality describes the theory for commutative groups, as a generalised Fourier transform.

Lie groups — Many important Lie groups are compact, so the results of compact representation theory apply to them. Other techniques specific to Lie groups are used as well. Most of the groups important in physics and chemistry are Lie groups, and their representation theory is crucial to the application of group theory in those fields. See Representations of Lie groups and Representations of Lie algebras.

Linear algebraic groups (or more generally affine group schemes) — These are the analogues of Lie groups, but over more general fields than just R or C. Although linear algebraic groups have a classification that is very similar to that of Lie groups, and give rise to the same families of Lie algebras, their representations are rather different (and much less well understood). The analytic techniques used for studying Lie groups must be replaced by techniques from algebraic geometry, where the relatively weak Zariski topology causes many technical complications.

Non-compact topological groups — The class of non-compact groups is too broad to construct any general representation theory, but specific special cases have been studied, sometimes using ad hoc techniques. The semisimple Lie groups have a deep theory, building on the compact case. The complementary solvable Lie groups cannot be classified in the same way. The general theory for Lie groups deals with semidirect products of the two types, by means of general results called Mackey theory, which is a generalization of Wigner's classification methods.

Representation theory also depends heavily on the type of vector space on which the group acts. One distinguishes between finite-dimensional representations and infinite-dimensional ones. In the infinite-dimensional case, additional structures are important (e.g. whether or not the space is a Hilbert space, Banach space, etc.).

One must also consider the type of field over which the vector space is defined. The most important case is the field of complex numbers. The other important cases are the field of real numbers, finite fields, and fields of p-adic numbers. In general, algebraically closed fields are easier to handle than non-algebraically closed ones. The characteristic of the field is also significant; many theorems for finite groups depend on the characteristic of the field not dividing the order of the group.

Reducibility

A subspace W of V that is invariant under the group action is called a subrepresentation. If V has exactly two subrepresentations, namely the zero-dimensional subspace and V itself, then the representation is said to be irreducible; if it has a proper subrepresentation of nonzero dimension, the representation is said to be reducible. The representation of dimension zero is considered to be neither reducible nor irreducible, just as the number 1 is considered to be neither composite nor prime.

Under the assumption that the characteristic of the field K does not divide the size of the group, representations of finite groups can be decomposed into a direct sum of irreducible subrepresentations (see Maschke's theorem). This holds in particular for any representation of a finite group over the complex numbers, since the characteristic of the complex numbers is zero, which never divides the size of a group.

In the example above, the first two representations given (? and s) are both decomposable into two 1-dimensional subrepresentations (given by span{(1,0)} and span{(0,1)}), while the third representation (t) is irreducible.

Generalizations

Set-theoretical representations

A set-theoretic representation (also known as a group action or permutation representation) of a group G on a set X is given by a function ? : G ? XX, the set of functions from X to X, such that for all g1, g2 in G and all x in X:

$$\rho(1)[x] = x\rho(g_1g_2)[x] = \rho(g_1)[\rho(g_2)[x]],$$

where 1 is the identity element of G. This condition and the axioms for a group imply that (g) is a bijection (or permutation) for all g in G. Thus we may equivalently define a permutation representation to be a group homomorphism from G to the symmetric group SX of X.

For more information on this topic see the article on group action.

Representations in other categories

Every group G can be viewed as a category with a single object; morphisms in this category are just the elements of G. Given an arbitrary category C, a representation of G in C is a functor from G to C. Such a functor selects an object X in C and a group homomorphism from G to Aut(X), the automorphism group of X.

In the case where C is VectK, the category of vector spaces over a field K, this definition is equivalent to a linear representation. Likewise, a set-theoretic representation is just a representation of G in the category of sets.

When C is Ab, the category of abelian groups, the objects obtained are called G-modules.

For another example consider the category of topological spaces, Top. Representations in Top are homomorphisms from G to the homeomorphism group of a topological space X.

Two types of representations closely related to linear representations are:

projective representations: in the category of projective spaces. These can be described as "linear representations up to scalar transformations".

affine representations: in the category of affine spaces. For example, the Euclidean group acts affinely upon Euclidean space.

BIBLIOGRAPHY

- Allenby, R. B. J. T. (1991), Rings, Fields and Groups (Second ed.), Edward Arnold, London, p. xxvi+383, ISBN 0-7131-3476-3, MR 1144518
- Bagaria, Joan (2020), Set Theory, in Zalta, Edward N. (ed.), The Stanford Encyclopedia of Philosophy (Spring 2020 ed.), Metaphysics Research Lab, Stanford University, retrieved 2020-08-20
- Bishop, Errett (1967), Foundations of Constructive Analysis, New York: Academic Press, ISBN 4-87187-714-0
- Blyth, T.S.; Robertson, E.F. (1985), Groups, Rings and Fields: Algebra through practice, Book 3, Cambridge: Cambridge University Press, ISBN 0-521-27288-2
- Bolzano, Bernard (1975), Berg, Jan (ed.), Einleitung zur Größenlehre und erste Begriffe der allgemeinen Größenlehre, Bernard-Bolzano-Gesamtausgabe, edited by Eduard Winter et al., vol. II, A, 7, Stuttgart, Bad Cannstatt: Friedrich Frommann Verlag, p. 152, ISBN 3-7728-0466-7
- Bourbaki, Nicolas, Commutative algebra. Chapters 1--7. Translated from the French. Reprint of the 1989 English translation. Elements of Mathematics (Berlin). Springer-Verlag, Berlin, 1998. xxiv+625 pp. ISBN 3-540-64239-0
- Bourbaki, Nicolas, Éléments de mathématique. Algèbre commutative. Chapitres 8 et 9. (Elements of mathematics. Commutative algebra. Chapters 8 and 9) Reprint of the 1983 original. Springer, Berlin, 2006. ii+200 pp. ISBN 978-3-540-33942-7

- Cantone, Domenico; Ferro, Alfredo; Omodeo, Eugenio G. (1989), Computable Set Theory, International Series of Monographs on Computer Science, Oxford Science Publications, Oxford, UK: Clarendon Press, pp. xii, 347, ISBN 0-198-53807-3
- Dauben, Joseph (1979), Georg Cantor: His Mathematics and Philosophy of the Infinite, Harvard University Press, pp. 30–54, ISBN 0-674-34871-0.
- Eisenbud, David (1995). Commutative algebra with a view toward algebraic geometry. Graduate Texts in Mathematics. Vol. 150. New York: Springer-Verlag. xvi+785. ISBN 0-387-94268-8. MR 1322960.
- Ernst Kunz, Introduction to Commutative algebra and algebraic geometry, Birkhauser 1985, ISBN 0-8176-3065-1
- Faith, Carl (1999), Rings and Things and a Fine Array of Twentieth Century Associative Algebra, Mathematical Surveys and Monographs, vol. 65, Providence, RI: American Mathematical Society, ISBN 0-8218-0993-8, MR 1657671
- Feferman, Solomon (1998), In the Light of Logic, New York: Oxford University Press, pp. 280–283, 293–294, ISBN 0-195-08030-0
- Ferro, Alfredo; Omodeo, Eugenio G.; Schwartz, Jacob T. (September 1980), Decision Procedures for Elementary Sublanguages of Set Theory. I. Multi-Level Syllogistic and Some Extensions, Communications on Pure and Applied Mathematics, 33 (5): 599–608, doi:10.1002/cpa.3160330503
- Forster, T. E. (2008). The iterative conception of set (PDF). The Review of Symbolic Logic. 1: 97–110.
- Frank Ruda (6 October 2011). Hegel's Rabble: An Investigation into Hegel's Philosophy of Right. Bloomsbury Publishing. p. 151. ISBN 978-1-4411-7413-0.
- Goodearl, K. R.; Warfield, R. B., Jr. (1989), An Introduction to Noncommutative Noetherian Rings, London Mathematical Society Student Texts, vol. 16, Cambridge: Cambridge University Press, ISBN 0-521-36086-2, MR 1020298
- homotopy type theory in nLab
- Homotopy Type Theory: Univalent Foundations of Mathematics. The Univalent Foundations Program. Institute for Advanced Study.
- Introduction to Sets. www.mathsisfun.com. Retrieved 2020-08-20.

- Jean-Pierre Serre, Local algebra. Translated from the French by CheeWhye Chin and revised by the author. (Original title: Algèbre locale, multiplicités) Springer Monographs in Mathematics. Springer-Verlag, Berlin, 2000. xiv+128 pp. ISBN 3-540-66641-9
- Jech, Thomas (2003), Set Theory, Springer Monographs in Mathematics (Third Millennium ed.), Berlin, New York: Springer-Verlag, p. 642, ISBN 978-3-540-44085-7, Zbl 1007.03002
- Johnson, Philip (1972), A History of Set Theory, Prindle, Weber & Schmidt, ISBN 0-87150-154-6
- Judson, Thomas W. (1997), Abstract Algebra: Theory and Applications
- Kimberling, Clark (1981), Emmy Noether and Her Influence, in Brewer, James W; Smith, Martha K (eds.), Emmy Noether: A Tribute to Her Life and Work, Marcel Dekker, pp. 3–61
- Kolmogorov, A.N.; Fomin, S.V. (1970), Introductory Real Analysis (Rev. English ed.), New York: Dover Publications, pp. 2–3, ISBN 0486612260, OCLC 1527264
- Lam, T. Y. (1999), Lectures on Modules and Rings, Graduate Texts in Mathematics, vol. 189, New York: Springer-Verlag, doi:10.1007/978-1-4612-0525-8, ISBN 0-387-98428-3, MR 1653294
- Lam, T. Y. (2001), A First Course in Noncommutative Rings, Graduate Texts in Mathematics, vol. 131 (Second ed.), New York: Springer-Verlag, doi:10.1007/978-1-4419-8616-0, ISBN 0-387-95183-0, MR 1838439
- Lam, T. Y. (2003), Exercises in Classical Ring Theory, Problem Books in Mathematics (Second ed.), New York: Springer-Verlag, ISBN 0-387-00500-5, MR 2003255
- Mac Lane, Saunders; Moerdijk, leke (1992), Sheaves in Geometry and Logic: A First Introduction to Topos Theory, Springer-Verlag, ISBN 978-0-387-97710-2
- Matsumura, Hideyuki (1989), Commutative Ring Theory, Cambridge Studies in Advanced Mathematics, vol. 8 (Second ed.), Cambridge, UK.: Cambridge University Press, ISBN 0-521-36764-6, MR 1011461
- Matsumura, Hideyuki, Commutative algebra. Second edition. Mathematics Lecture Note Series, 56. Benjamin/Cummings Publishing Co., Inc., Reading, Mass., 1980. xv+313 pp. ISBN 0-8053-7026-9
- Matsumura, Hideyuki, Commutative Ring Theory. Second edition. Translated from the Japanese. Cambridge Studies in Advanced Mathematics, Cambridge, UK : Cambridge University Press, 1989. ISBN 0-521-36764-6

- McConnell, J. C.; Robson, J. C. (2001), Noncommutative Noetherian Rings, Graduate Studies in Mathematics, vol. 30, Providence, RI: American Mathematical Society, doi:10.1090/gsm/030, ISBN 0-8218-2169-5, MR 1811901
- Michael Atiyah & Ian G. Macdonald, Introduction to Commutative Algebra, Massachusetts : Addison-Wesley Publishing, 1969.
- Miles Reid, Undergraduate Commutative Algebra (London Mathematical Society Student Texts), Cambridge, UK : Cambridge University Press, 1996.
- Nagata, Masayoshi, Local rings. Interscience Tracts in Pure and Applied Mathematics, No. 13. Interscience Publishers a division of John Wiley and Sons, New York-London 1962 xiii+234 pp.
- O'Connor, J. J.; Robertson, E. F. (September 2004), The development of ring theory, MacTutor History of Mathematics Archive
- Pierce, Richard S. (1982), Associative Algebras, Graduate Texts in Mathematics, vol. 88, New York: Springer-Verlag, ISBN 0-387-90693-2, MR 0674652
- Rémi Goblot, Algèbre commutative, cours et exercices corrigés, 2e édition, Dunod 2001, ISBN 2-10-005779-0
- Rodych 2018, §2.1: When we prove a theorem or decide a proposition, we operate in a purely formal, syntactical manner. In doing mathematics, we do not discover pre-existing truths that were 'already there without one knowing' (PG 481)—we invent mathematics, bit-by-little-bit. Note, however, that Wittgenstein does not identify such deduction with philosophical logic; c.f. Rodych §1, paras. 7-12.
- Rodych 2018, §2.2: An expression quantifying over an infinite domain is never a meaningful proposition, not even when we have proved, for instance, that a particular number n has a particular property. Rodych 2018, §3.6.
- Rodych 2018, §3.4: Given that mathematics is a 'motley of techniques of proof' (RFM III, §46), it does not require a foundation (RFM VII, §16) and it cannot be given a self-evident foundation (PR §160; WVC 34 & 62; RFM IV, §3). Since set theory was invented to provide mathematics with a foundation, it is, minimally, unnecessary.
- Rodych, Victor (Jan 31, 2018). Wittgenstein's Philosophy of Mathematics. In Zalta, Edward N. (ed.). Stanford Encyclopedia of Philosophy (Spring 2018 ed.).

INDEX